Postcranial anatomy and habits of Asian multituberculate mammals

ZOFIA KIELAN-JAWOROWSKA AND PETR P. GAMBARYAN

T0225981

Kielan-Jaworowska, Z. & Gambaryan, P.P. 1994 12 15: Postcranial anatomy and habits of Asian multituberculate mammals. *Fossils and Strata*, No. 36, pp. 1–92. Oslo. ISSN 0300–9491. ISBN 82-00-37650-8.

Postcranial skeletal elements of six Late Cretaceous taeniolabidoid taxa from the Gobi Desert are described, and the postcranial musculature of *Kryptobaatar* and *Nemegtbaatar* is reconstructed. A new reconstruction of multituberculate pes is given, showing Mt III abducted 30° from the longitudinal axis of the tuber calcanei. The calcaneo–Mt V contact and abduction–adduction at the astragalonavicular joint in a horizontal plane, around a vertical axis, are recognized as new multituberculate autapomorphies. Other new, partly plesiomorphic, multituberculate characters are: no transverse foramen in atlas; cervical ribs present at least in some taxa; large iliosacral angle (35–37°); iliosacral contact dorsoventral rather than mediolateral; an incipient supraspinous fossa and a peg-like acromion. The deep multituberculate pelvis with femoral adductors originating ventral to the acetabulum and the large mediolateral diameter of the tibia indicate abduction of the femora by 30–60°, while twisting of the humerus is indicative of abducted forelimbs. Long spinous processes of the lumbar vertebrae, sloping craniodorsally in Asian multituberculates, suggest asymmetrical gait and long jumps, but the short tibia and short Mt III suggest short jumps. This inconsistency is due to the abducted limbs, because of which the direct transposition of jump mechanics of mammals with parasagittal limbs does not work for multituberculates. Multituberculates possibly had a steeper trajectory of jump than modern therian mammals. The studied Asian multituberculates do not show adaptations to arborealism. It is suggested that the coracosternal joint disappeared in multituberculates (and independently in therians) as an adaptation to asymmetrical gait. It is speculated that the competitive inferiority of multituberculates to eutherians is related to the structure of the pelvis with a ventral keel, which hindered prolongation of the gestation period, and to abducted limbs that limited their endurance for prolonged running. The analysis of multiturbeculate–therian postcranial synapomorphies does not support the idea that Multituberculata is the sister taxon of Theria. □*Mammalia, Multituberculata, Anatomy, Cretaceous, Asia.*

Zofia Kielan-Jaworowska, Paleontologisk Museum, Universitetet i Oslo, Sars Gate 1, N–0562 Oslo 5, Norway; Petr P. Gambaryan, Zoological Institute, Russian Academy of Sciences, Universitetskaya Naberezhnaya 1, 199164 St. Petersburg, Russia; 27th September, 1993; revised 21st April, 1994.

Contents

Introduction 3
    Terminology 5
Osteological descriptions 6
    Suborder Cimolodonta McKenna, 1975 (new rank); Infraorder
        Taeniolabidoidea Sloan & Van Valen, 1965 (new rank);
        Family Eucosmodontidae Jepsen, 1940 6
Genus *Kryptobaatar* Kielan-Jaworowska, 1970 6
Kryptobaatar dashzevegi Kielan-Jaworowska, 1970 6
    Skull 6
    Axial skeleton 6
    Pectoral girdle (forelimb unknown) 7
    Pelvic girdle and hind limb 7

Genus *Nemegtbaatar* Kielan-Jaworowska, 1974 17
Nemegtbaatar gobiensis Kielan-Jaworowska, 1974 17
    Skull 17
    Axial skeleton 17
    Pectoral girdle and forelimb 22
    Pelvis and hind limb 26
Genus *Chulsanbaatar* Kielan-Jaworowska, 1974 30
Chulsanbaatar vulgaris Kielan-Jaworowska, 1974 30
    Skull 31
    Hyoid apparatus 31
    Axial skeleton 32
    Pectoral girdle and forelimb 36

Pelvic girdle and hind limb 36
Family Sloanbaataridae Kielan-Jaworowska, 1974 38
Genus *Sloanbaatar* Kielan-Jaworowska, 1970 38
Sloanbaatar mirabilis Kielan-Jaworowska, 1970 38
    Axial skeleton 38
    Pelvic girdle and hind limb 38
Family Taeniolabididae Granger & Simpson, 1929 39
Genus *Catopsbaatar* Kielan-Jaworowska, 1994 39
Catopsbaatar catopsaloides (Kielan-Jaworowska, 1974) 39
Taeniolabidoid, fam. gen. et sp. indet. (Kielan-Jaworowska 1989) ... 40
    Axial skeleton 40
    Pectoral girdle 40
Myological reconstructions 40
Muscles of the forelimb 40
Muscles of the axial skeleton 45
Muscles of the pelvic girdle and hind limb 47

Anatomical comparisons 57
Proportions of the body 57
Hyoid apparatus 57
Vertebral column 57
Pectoral girdle and forelimb 60
Pelvic girdle and hind limb 62
Functional anatomy 65
Reconstruction of locomotion 65
Structure and function of multituberculate pes 74
Pedal adaptations of Asian and North American
    multituberculates 79
Forelimb movements 82
Concluding remarks 82
Plesiomorphies and apomorphies of multituberculates 83
Habits and extinction 85
References 88

Introduction

Remains of the Multituberculata (assigned to a subclass of their own, Allotheria Marsh, 1880) are found in deposits that range in age from the Late Triassic (Rhaetian; Sigogneau-Russell 1989) through Late Eocene (recognized previously as Early Oligocene by Krishtalka *et al.* 1982, but see Swisher & Prothero 1990). Although multituberculates are the most common fossils in the majority of well sampled Late Cretaceous and Paleocene localities in the Northern Hemisphere, their postcranial remains are rarely found and are still incompletely known.

Multituberculate postcranial elements or partial skeletons have been found only in members of the Cimolodonta McKenna, 1975 (a taxon erected as an infraorder to include the 'parvorders' Ptilodontoidea Sloan & Van Valen, 1965, and Taeniolabidoidea Sloan & Van Valen, 1965, the latter two regarded by us as infraorders). The cimolodont postcranial fragments from the Paleocene or, less often, Late Cretaceous, were described or discussed in the 20th century by Gidley (1909), Broom (1914), Simpson (1926, 1928a, 1937), Simpson & Elftman (1928), Granger & Simpson (1929), McKenna (1961), Clemens (1963), Deischl (1964), Sloan & Van Valen (1965), Kielan-Jaworowska (1969, 1979, 1989), Sahni (1972), Jenkins (1973), Kielan-Jaworowska & Dashzeveg (1978), Jenkins & Weijs (1979), Krause & Baird (1979), Krause & Jenkins (1983), Jenkins & Krause (1983), Kielan-Jaworowska & Qi (1990), Sereno & McKenna (1990), Kielan-Jaworowska & Nessov (1992) and Szalay (1993). The most comprehensive work to date is the paper by Krause & Jenkins (1983), which includes the description of a fairly complete postcranial skeleton of the Paleocene genus *Ptilodus*, as well as a thorough review of multituberculate postcranial literature, including 19th century papers of Cope and Marsh, not cited by us.

The aim of this monograph is to describe multituberculate postcranial material from the Late Cretaceous Djadokhta and Barun Goyot formations (and their stratigraphic equivalents) in the Gobi Desert, assembled by the Polish–Mongolian Palaeontological Expeditions, and to reconstruct musculature and habits of the animals. The exact age of these formations remains an open question (e.g., Gradziński *et al.* 1977; Fox 1978; Lillegraven & McKenna 1986). We tentatively follow the estimates given by Gradziński *et al.* (1977): the Djadokhta Formation (and its stratigraphic equivalent the Toogreeg beds) is regarded as upper Santonian and/or lower Campanian; the Barun Goyot Formation (and its stratigraphic equivalent the Red beds of Khermeen Tsav) as middle Campanian.

The most complete postcranial skeletons described here belong to *Kryptobaatar dashzevegi* Kielan-Jaworowska, 1970, *Nemegtbaatar gobiensis* Kielan-Jaworowska, 1974, and *Chulsanbaatar vulgaris* Kielan-Jaworowska, 1974; less complete postcranial fragments of *Sloanbaatar mirabilis* Kielan-Jaworowska, 1970, *Catopsbaatar catopsaloides* (Kielan-Jaworowska, 1974) and a taeniolabidoid, fam., gen. et sp. indet. (Kielan-Jaworowska 1989), are also briefly described (see Kielan-Jaworowska 1970, 1971, 1974, 1994; Kielan-Jaworowska & Sloan 1979; Kielan-Jaworowska *et al.* 1986; and Hurum 1992, 1994, for description of the skulls of these taxa). All Asian Late Cretaceous multituberculates belong to the Taeniolabidoidea. We describe the postcranial skeletons of the above-mentioned taxa, beginning with those that are more complete or better preserved. As the taeniolabidoid postcranial skeleton is fairly uniform morphologically, the description of elements that do not differ from those better preserved in other taxa is omitted.

In contrast to the North American Late Cretaceous, where the postcranial mammalian skeletons are preserved as isolated elements (e.g., Clemens 1963; McKenna 1961; Deischl 1964; Sloan & Van Valen 1965; Szalay & Decker 1974; Sahni 1972; Szalay 1984, 1993; Krause & Baird 1979; Krause & Jenkins 1983; Bleefeld 1992; see also Clemens & Kielan-Jaworowska 1979 for review), the Late Cretaceous Gobi

Desert postcranials occur as more or less complete articulated skeletons. This has its obvious advantages, but also some drawbacks. Because of the scarcity of the material (e.g., the postcranial skeleton of *Kryptobaatar* is known from a single specimen) and the minute size of most of the taxa, it proved impossible to separate the particular bones and examine them from all sides. Consequently, the articular surfaces in most cases could not be studied. For comparative purposes we studied isolated postcranial elements of the Late Cretaceous North American multituberculates, a partial skeleton of *?Eucosmodon* sp. from the Paleocene of North America and humeri of *?Lambdopsalis bulla* Chow & Qi, 1978, from the Eocene of China, in which articular surfaces, and sometimes also muscle scars, are well preserved.

Reconstructions of the musculature of fossil mammals have rarely been attempted, although there is a wealth of skeletal remains. The difficulties and reliability of muscle reconstructions in fossil vertebrates were recently summarized by Bryant & Seymour (1990) and Bleefeld (1992). In spite of these difficulties, we believe that in cases when bone surface is excellently preserved, relatively reliable muscle reconstructions in fossil mammals are possible.

The masticatory apparatus and jaw movements of multituberculates were reconstructed by Simpson (1926) for the Late Jurassic *Ctenacodon*, by Sloan (1979) for the Paleocene *Ectypodus*, and by Krause (1982) and Wall & Krause (1992) for the Paleocene *Ptilodus* (see also Broom 1910; Simpson 1933; Turnbull 1970; Gingerich 1977, 1984; Hopson *et al.* 1989, for discussion of multituberculate jaw mechanics). As far as the postcranial skeleton is concerned, Simpson & Elftman (1928) reconstructed the hind-limb musculature of *?Eucosmodon* from the Paleocene of North America, while Jenkins & Krause (1983) and Krause & Jenkins (1983) discussed climbing specializations of *Ptilodus* and *?Eucosmodon*. The musculature of Cretaceous multituberculates has not been reconstructed as yet.

We reconstruct the postcranial musculature of *Kryptobaatar* and *Nemegtbaatar* and some muscles of *Chulsanbaatar*, *Sloanbaatar*, an unidentified taeniolabidoid from the Djadokhta Formation, and *Lambdopsalis*. Although the neck vertebrae in *Nemegtbaatar* and *Chulsanbaatar* have been preserved, it was impossible to reconstruct the neck muscles because the relevant muscle scars were not discernible.

We have studied in detail and figured the skeletons and musculature of several extant, small, terrestrial marsupials and rodents: the marsupial *Antechinus stuarti* and the rodents *Meriones blackleri*, *Meriones tamariscinus* and *Mesocricetus branti*. In addition, many other extant species that are not figured have been dissected, measured and studied (Tables 1–6). Comparisons with the musculature of monotremes was less useful, because of their locomotor specializations. All extant monotremes (*Ornithorhynchus*, *Tachyglossus* and *Zaglossus*) are powerful diggers, and in addition *Ornithorhynchus* is adapted to a semi-aquatic mode of life. In contrast, small marsupial and eutherian mammals appear to share ecological specializations with the multituberculates presented here.

The muscle scars (especially of small muscles) preserved on the bones of the studied multituberculates are not always obvious. As multituberculates differ from extant marsupial and eutherian mammals in many important details of their cranial and postcranial anatomy, it is possible that some muscles present in extant mammals used by us for comparisons were not present in multituberculates and vice versa. For these reasons, our reconstructions of multituberculate musculature should be considered as best approximations.

In order to understand the position and movements of multituberculate hind limbs, a plastic model of an idealized pelvis and hind limbs (based on *Kryptobaatar* and *Nemegtbaatar*) has been made. Because of the incompleteness of the material, it was impossible to make such a model of the forelimbs and shoulder girdle. Our reconstructions of multituberculate hind-limb movements during the propulsive phase (Figs. 51–53) are partly based on this model.

Origins of multituberculates and their relationships to other mammals are debated (see, e.g., Crompton & Jenkins 1973; Hahn 1973; Jenkins 1984; Hopson & Barghusen 1986; Kemp 1982, 1983; Rowe 1988, 1993; Szalay 1990, 1993; Wible 1991; Kielan-Jaworowska 1992; Miao 1991, 1993; and Wible & Hopson 1993 for recent reviews). The problem of the origins and relationships of multituberculates is beyond the scope of the present monograph. However, we hope that the recognition of multituberculate plesiomorphies and apomorphies based on the postcranial anatomy may bring new data for understanding the relationships of multituberculates to other mammals.

Acknowledgments. – During the work on this monograph we greatly benefitted from discussion and correspondence with numerous colleagues. We are especially grateful to David W. Krause, Kenneth D. Rose and Fred S. Szalay who read the whole manuscript and offered most useful criticism and suggestions. We appreciate also the assistance of other colleagues who read parts of the manuscript and commented on it. Prime among them were: Ann R. Bleefeld, Percy M. Butler, Jean-Pierre Gasc, Farish A. Jenkins, Jr., Francoise K. Jouffroy, Alexander N. Kuznetsov, Denise Sigogneau-Russell and Hans-Dieter Sues. Malcolm C. McKenna (American Museum of Natural History, New York) and Qi Tao (Institute of Vertebrate Paleontology and Paleoanthropology, Beijing) kindly allowed us to study the specimens in their charge. The specimens described by us were prepared during several years by the members of the technical staff of the Institute of Paleobiology of the Polish Academy of Sciences in Warsaw. Anna Ohandganian-Gambaryan (St. Petersburg) generously offered her time for various kinds of technical assistance. The photographs were taken partly at the Institute of Paleobiology in Warsaw by the late Maria Czarnocka and by Elzbieta Mulawa and Marian Dziewinski, the remainder at the Palaeontological Museum, Oslo by Per Ås. The drawings have been inked by Evgenii A. Bessonov, Natalia A. Florenskaya and Galina E. Zubtsova in St. Petersburg. Jørn Hurum (Palaeontological Museum, Oslo) helped us in computer editing of the drawings. The work of ZKJ has been supported by Norges Allmenvitenskapelige Forskningsråd, grants Nos. 441.91/002, 441.92/003 and 441.93/001. The work of PPG has been supported by the Russian Academy of Sciences, grant no. 93-04-07699. PPG received from the Polish Academy of Sciences a fellowship to visit Warsaw in 1979, when our cooperation started. The cooperation was subsequently postponed for several years and then renewed in 1991 when PPG received a grant from Norges Allmenvitenskapelige Forskningsråd to visit Oslo and then a fellowship for stay in Oslo from December 1992 to March 1993. The visits of ZKJ in St. Petersburg in May 1993 and of PPG in Oslo in September 1993 were supported in part by the University of Oslo. The publication of this monograph is funded by Norges Allmenvitenskapelige Forskningsråd, grant No. 441.93/004. To all these persons and institutions we would like to express our sincere thanks and gratitude.

Terminology

Until recently, Monotremata were assigned to the subclass Prototheria Gill, 1872, and Theria Parker & Haswell (1897) included Marsupialia, Eutheria and extinct groups traditionally classified as Symmetrodonta and Eupantotheria (but see McKenna 1975; Szalay 1977; Rowe 1988; Novacek 1992, for attempts of a modern classification of Mammalia and Eutheria). It may be inferred from the discovery of a Cretaceous monotreme, *Steropodon*, that the Monotremata might be an early offshoot of the Theria (Archer *et al.* 1985; Kielan-Jaworowska *et al.* 1987; Jenkins 1990; Kielan-Jaworowska 1992). In spite of this, in the present monograph we refer to the extant marsupials and eutherians in the traditional way as therian mammals (i.e. excluding the Monotremata).

We were unable to follow the osteological terminology of *Nomina Anatomica Veterinaria* (International Commission on Veterinary Anatomical Nomenclature 1973, see also Schaller 1992), since our terminology would then differ from that of the major reviews of vertebrate paleontology (e.g., Gregory 1910, 1951; Romer 1966; Romer & Parsons 1986; Carroll 1988; Benton 1990 and many others) and from papers on multituberculate postcranial anatomy (e.g., Gidley 1909; Granger & Simpson 1929; Deischl 1964; Krause & Jenkins 1983; Jenkins & Krause 1983). In these textbooks and papers, terms such as the *astragalus, navicular, cuboid,* etc., are generally used instead of *talus, os tarsi centrale, os tarsale IV,* etc. In an attempt to adhere to standard terminology we follow that of Krause & Jenkins (1983). When an anatomical term differs from that in the *Nomina Anatomica Veterinaria,* its other names are given in brackets at its first mention. The paper by Krause & Jenkins (1983) is followed in referring to the *dorsal* and *ventral* aspects of the humerus and femur rather than 'anterior' and 'posterior', as this better reflects the anatomical position of the bones, not only in multituberculates but in all Mesozoic and many extant, especially nidicolous, mammals.

For myological terminology *Nomina Anatomica Veterinaria* (1973) is followed in cases when the muscles discussed occur in domestic mammals. For other muscles we use mostly Elftman (1929), Howell (1933), Rinker (1954) and Jouffroy (1971).

The terminology of the gait is confusing. We follow Jenkins & Goslow (1983) using *propulsive phase* (rather than 'phase of support' of Howell 1944 and Gambaryan 1974) when the foot (or feet) strikes the ground. In galloping mammals there may occur two phases when all four feet are clear of the ground. Muybridge (1877, cited here from the

1957 edition) did not give formal names to the two phases, but he spoke about the 'flight' with outstretched legs and the 'spring' with legs flexed under the body. The two phases together have been referred to as 'unsupported period' (Hildebrand 1960) or 'free flight' (Gambaryan 1974). Howell (1944) called the phase with extended legs the 'centrifugal suspension' and the one with gathered legs 'centripetal suspension'. The same phases have been referred to respectively as 'extended' and 'bunched suspension' (Brown & Yalden 1973), 'extended' and 'gathered suspension' (Hildebrand 1988), and 'extended' and 'crossed flight' (Gambaryan 1974). 'Free flight' of Gambaryan (1974) is a translation of the term 'фаза свободного полёта' used in Russian zoological literature. It seems to us that 'flight' better portrays the inertial movement of an animal in the air than 'suspension', which usually refers to a static state of a hanging body or of particulates in a fluid or gas. We therefore prefer the term *flight,* but the two phases of flight are referred to as *extended flight* and *gathered flight* (rather than 'crossed flight'), as again these terms better portray the posture of the animal.

For brevity, only the generic names are usually used in the descriptions and discussions, since all the Mongolian fossil taxa described belong to monotypic genera. A synonymy list is given only for *Catopsbaatar catopsaloides,* the generic assignment of which has been changed. Under 'Material' we cite only the specimens in which the postcranial fragments have been preserved.

Two new terms are introduced:

- a tubercle that in dorsal aspect of the femur is placed at the divergence of the neck and the greater trochanter, is called the *subtrochanteric tubercle* (Figs. 3A, 16B, 21D, 24B, C);

- a fissure-like fossa, placed lateral to the lesser trochanter and referred to by Simpson & Elftman (1928) as a part of the divided trochanteric (digital) fossa, is called the *posttrochanteric fossa* (Fig. 16A).

The following abbreviations for museum collections and localities are used: AMNH – American Museum of Natural History, New York; IVPP – Institute of Vertebrate Paleontology and Paleoanthropology, Beijing; UA – University of Alberta, Edmonton; UCMP – Museum of Paleontology, University of California, Berkeley; ZPAL – Institute of Paleobiology, Polish Academy of Sciences, Warsaw; ZIN – Zoological Institute, Russian Academy of Sciences, St. Petersburg.

Other abbreviations are: abd. = abductor, add. = adductor, c. = caput, C = cervical vertebrae, Cd = caudal vertebrae, cond. = condylus; D I – D V = digits I–V, ext. = extensor, f. = foramen, fac. = facies, fl. = flexor, inc. = incisura, L = lumbar vertebrae, lig. = ligamentum, m. = muscle, Mt I – Mt V = metatarsals I–V, p. = pars, Ph 1–3 = phalanges 1–3, proc. = processus, S = sacral vertebrae, T = thoracic vertebrae, troch. = trochanter.

Osteological descriptions

Suborder Cimolodonta McKenna, 1975 (new rank)

Infraorder Taeniolabidoidea Sloan & Van Valen, 1965 (new rank)

Family Eucosmodontidae Jepsen, 1940

Genus *Kryptobaatar* Kielan-Jaworowska, 1970

Synonym. – Gobibaatar Kielan-Jaworowska, 1970

Kryptobaatar dashzevegi Kielan-Jaworowska, 1970

(Figs. 1–3, 5–7, 37D, 44A, 54, 59A)

Material. – Djadokhta Formation, Bayn Dzak, about 1 km eastward from the Main Field, ZPAL MgM-I/41 (the specimen was found *in situ*): damaged skull and partial right dentary, badly damaged axis and left side of C3–C5, incomplete right scapulocoracoid, and in anatomical succession incomplete L5, L6–L7, incomplete sacrum, anterior caudal vertebrae (Cd1–Cd4; Cd4 fragmentary), complete right and left pelves with epipubic bones preserved on both sides, almost complete right and left hind limbs in anatomical arrangement, with femora displaced medially; one posterior caudal found in the same piece of rock.

Skull
Fig. 1

The postcranial skeleton of *Kryptobaatar* (ZPAL MgM-I/41) described below was found in association with the skull and partial right dentary, which have not yet been figured. The 'basioccipital box' of Kielan-Jaworowska & Dashzeveg (1978) (Fig. 1D) was subsequently recognized as an artifact (Kielan-Jaworowska *et al.* 1986). The basicranial region of *Kryptobaatar* does not differ from those of other taeniolabidoid multituberculate genera. The skull is very large with respect to the length of the body, its estimated length being about 34 mm, which is about 1.1 times the length of the pelvis.

Axial skeleton
Figs. 2, 3, 6

Cervical vertebrae (not figured). – The axis fragment consists of a damaged body with bases of the arch on both sides. The dens and postzygapophyses (caudal articular surface, zygapophysis caudalis, proc. articularis caudalis) are missing. The bases of the transverse processes are broken. The pedicles (preserved on both sides) are relatively long and stout, indicating the presence of stout laminae. C3–C5 are badly damaged; only fragments of the bodies with bases of the left transverse processes are preserved.

Lumbar vertebrae (Figs. 2, 3A, B, 6C, see also Fig. 4 for method of measurements of the vertebrae). – The prezygapophyses (cranial articular surface, zygapophysis cranialis, proc. articularis cranialis) and postzygapophyses are very large. The lateral walls of the pedicles are strongly concave. The spinous processes (broken), preserved on L6 and L7, arise from the whole length of the arch and are massive, on L7 directed upwards, only slightly directed cranially, on L6 more strongly inclined cranially. The preserved parts of the spinous processes show that they were very long, much longer than in *Meriones* (see Fig. 36A). The transverse processes are broken off; judging from the preserved parts they were wider and more massive on L6 than on L7. There is a median ventral crest (crista ventralis) on L5 and L6.

Sacrum (Figs. 2, 3A–C). – The preserved parts of the sacrum consist of S1–S2, a fragment of S3, and, some distance caudally, a tiny fragment of S4 in articulation with Cd1. As the sacrum is firmly fused to the pelvis, its surface for articulation with the latter cannot be examined.

The estimated length of the sacrum is about 16 mm, its maximal width across the auricular surfaces is about 7 mm. S1 is shorter than the last lumbar and bears robust transverse processes directed craniolaterally downwards. The prezygapophyses (in articulation with L7) are large and face dorsocaudally. The position of the first dorsal sacral foramen shows that S2 is much longer than S1. The spinous processes are broken, but it may be inferred from the long base of the process preserved on S1 (Fig. 3A, B) that the sacral spinous processes were apparently as large as those of the last lumbar vertebrae. The transverse processes of S2 are longer than those of S1. Both articulate with the ilia, but when the transverse process of the first vertebra is covered by the ilium dorsally, that of the second vertebra meets the medial side of the ilium and is possibly less strongly synostosed to it. The craniolateral corner of the first transverse sacral process is thickened to form a rounded inflation, the very margin of which protrudes somewhat laterally and below, beyond the ilium. The cranial margin of the second dorsal foramen is preserved on the right side. The pedicles of the fused arches of the two first vertebrae arise dorsally, the laminae being arranged horizontally, at a right angle to the pedicles. The intermediate sacral crest is developed as a continuous longitudinal crest at the edge of the pedicle and the lamina. The laminae, as seen in dorsal view (Fig. 3A, B) form a trapezoid-shaped surface that narrows caudally. The median sacral crest was probably highest at the cranial part of S1. A poorly preserved indentation is present in front of an incompletely preserved spinous process of S2. The preserved part of S4 consists of only a badly damaged fragment of the arch and a somewhat displaced postzygapophysis.

Caudal vertebrae. – Cd1–Cd3 (Figs. 2, 3B–D), in articulation with a fragment of S4, were originally preserved above the ischia. The length of Cd1, measured between pre- and post-zygapophyses is 4.3 mm, that of Cd2 4.3 mm. The transverse processes are large and arise from the whole length of the bodies. The most complete is the left transverse process of Cd2, which is directed ventrally and twisted craniolaterally (in Fig. 3B it is almost completely preserved; it has been slightly damaged during the preparation and appears shorter in Fig. 3C.). Both pre- and postzygapophyses are large, the cranial articular surface faces mediodorsally, the caudal lateroventrally. The pedicles are lower than in the lumbar vertebrae, and the laminae are arranged horizontally. The spinous processes are stout and very long, as those of the lumbar vertebrae (see above); they arise from the middle part of the laminae, that of Cd1 is directed dorsally, Cd2 and Cd3 are somewhat inclined and are extended longitudinally at the extremities. The ventral aspect of Cd3 is partly exposed; there is no ventral crest.

The single middle caudal vertebra (Fig. 3D) is exposed in ventral view; its body is 7.4 mm long. The transverse processes arise from the whole length of the body and are directed lateroventrally. There is a prominent ventral median crest, and the surfaces between it and the extremities of the transverse processes are strongly concave.

Pectoral girdle (forelimb unknown)

Scapulocoracoid. – The right scapulocoracoid preserved in ZPAL MgM-I/41 (Fig. 5B–C) consists of an incomplete blade and a glenoid fossa. The scapular and coracoid parts of the glenoid fossa are arranged at an angle of about 120° with respect to each other. The suture between the glenoid and scapular parts is not discernible. As the *Kryptobaatar* scapulocoracoid is less complete than that of *Nemegtbaatar* ZPAL MgM-I/81 (see description of *Nemegtbaatar* below and Figs. 12 and 13G–J), in the description that follows we compare it with that of *Nemegtbaatar*. It cannot be stated with any certainty whether the glenoid part was L-shaped, as is characteristic of *Nemegtbaatar*, but it is possible that the medially directed process was present. The arcuate ridge in the ventral part of the blade is not recognized. The cranial border is possibly less prominent than in *Nemegtbaatar*, which may be due to the poor state of preservation of the *Kryptobaatar* scapulocoracoid. The infraspinous fossa appears to widen dorsally, as in *Nemegtbaatar*. Since there is a shallow fossa cranial to the spine, it appears that *Kryptobaatar* possessed an incipient supraspinous fossa, which is, however, less obvious than in *Nemegtbaatar*. The preserved part of the subscapular fossa is convex.

Pelvic girdle and hind limb

Pelvis (Figs. 2, 3A, 5A, 37D). – The length of the pelvis, measured on the right side from the cranial margin of the ilium to the end of the ischial tuber is 31.6 mm; when measured to the most caudal prominence of the ischial arc it is 32.8 mm. The pelvic sutures appear to be synostosed, but some of them are discernible: between the ischium and the pubis (between the acetabulum and the obturator foramen) on the medial aspect on both sides, and between the ilium and ischium above the cranial margin of the acetabulum, at a point opposite the craniocaudal midpoint of the acetabulum. The area of the suture between the ischium and the pubis below the obturator foramen is broken on both sides. The suture between the ilium and pubis is not discernible.

The ilium is 23 mm long. The wing of the ilium is rod-like and reflected laterally at its cranial end. The estimated distance between the cranial ends of the ilia is 16 mm. The ilia are roughly oval in cross-section and expanded medially in the area of the auricular surface (facies auricularis). Extending cranially from the acetabulum along the dorsal side of the ilium, there is a weak ridge (much less prominent than in *Nemegtbaatar*) for the origin of the dorsal part of m. gluteus medius. In lateral view the ilium increases in depth from the acetabulum for about two thirds of its length to the caudal ventral iliac spine, which forms a prominent process. Cranially to the process the depth of the ilium decreases again. Along the ventral margin there are two concavities: in front and to the rear of the caudal ventral iliac spine. The short cranial margin of the ilium is gently rounded. The wing in front of the auricular surface is concave. The shape of the auricular surface is not known, it is relatively long, extending for 8 mm. The ventral part of the ilium adjacent to the acetabulum is relatively small, blade-like and slightly concave in lateral view. Although the suture between the ilium and the pubis is not preserved, there is an elevation (which may correspond to the suture) that extends from the acetabulum and ends with the iliopubic eminence on the ventral margin.

The acetabulum is 3.4 mm long; as is characteristic of the multituberculates it is broadly and deeply emarginated dorsally, the articulated femoral heads being completely exposed dorsocaudally. The cranial rim of the acetabulum is highly elevated, forming a rounded ridge. In front of the cranial rim a low, triangular protuberance extends for about 4 mm. There is a small pit below the protuberance, close to the acetabulum, in a place where in other mammals there is a tuberosity for the origin of m. rectus femoris. Caudally the acetabulum is surrounded by an elevated border. Below this border there is a deep crescent-shaped notch on the ischium, situated outside the acetabulum. The obturator foramen is situated below and somewhat caudal to the acetabulum. It is oval, elongated craniocaudally and 3.7 mm long.

The ischium, measured from the suture with the ilium to the end of the ischial tuber, is 8.6 mm long; measured to the most caudal prominence of the ischial arc it is 9.8 mm long. Its dorsal margin is strongly thickened, forming a rounded ridge. It is sharply recurved dorsocaudally and forms a very prominent ischial tuber. This, as originally preserved (Fig. 2)

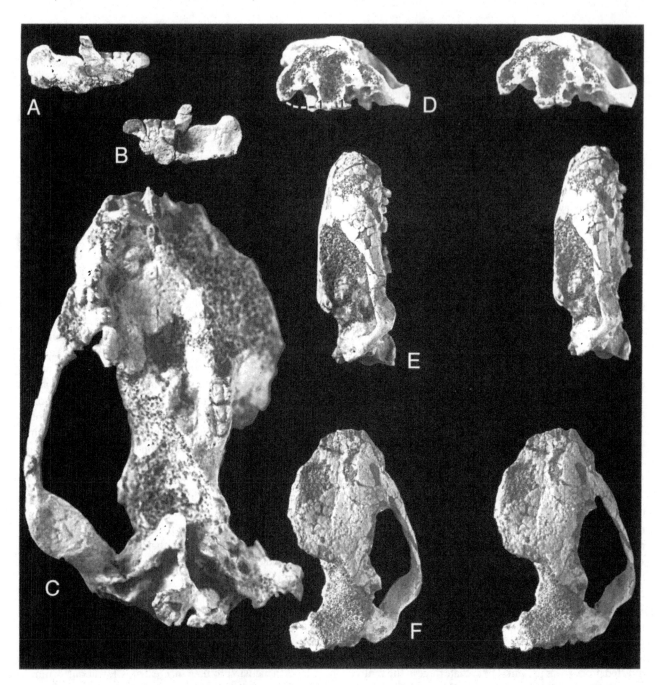

Fig. 1. Kryptobaatar dashzevegi (ZPAL MgM-I/41), Djadokhta Formation, Bayn Dzak, Gobi Desert, Mongolia. □A. Incomplete right dentary, lateral view. □B. Medial view of A. □C. Ventral view of incomplete skull of the same individual. □D, E, F. Same skull as in C, in occipital, right lateral and dorsal views. In D, on left side, missing ventral wall (dashed line) of presumable resonatory air space reconstructed. 1 = medial walls of air spaces forming the 'basioccipital box' of Kielan-Jaworowska & Dashzeveg (1978), recognized by Kielan-Jaworowska *et al.* (1986) as deformation artefact. A, B, D–F ×2; C ×4; D, E, F stereo-pairs.

Fig. 2 (opposite page). *Kryptobaatar dashzevegi* (ZPAL MgM-I/41), Djadokhta Formation, Bayn Dzak, Gobi Desert, Mongolia. Posterior part of skeleton found in association with skull and dentary in Fig. 1, consisting of: L5 (incomplete), L6, L7, damaged sacrum, Cd1–Cd3, almost complete hind limbs. □A. Right lateral view. □B. Left lateral view. 1 = incomplete L5; 2 = L6; 3 = L7; 4 = sacrum, preserved in two parts; 5 = Cd1; 6 = postobturator notch; 7 = calcaneum, on left side only the tuber calcanei has been preserved; 8 = parafibula; 9 = astragalus; 10 = cuboid; 11 = groove for tendon of m. peroneus longus; 12 = hook-like process of fibula; 13 = obturator foramen; 14 = epipubic bone; 15 = greater trochanter; 16 = lesser trochanter; arrow in A and upper arrow in B = left ischial tuber, right one not preserved; lower arrow in B = lateral malleolus of the fibula. Bone between distal ends of tibia and fibula in A possibly fragment of dentary. Both ×3.8.

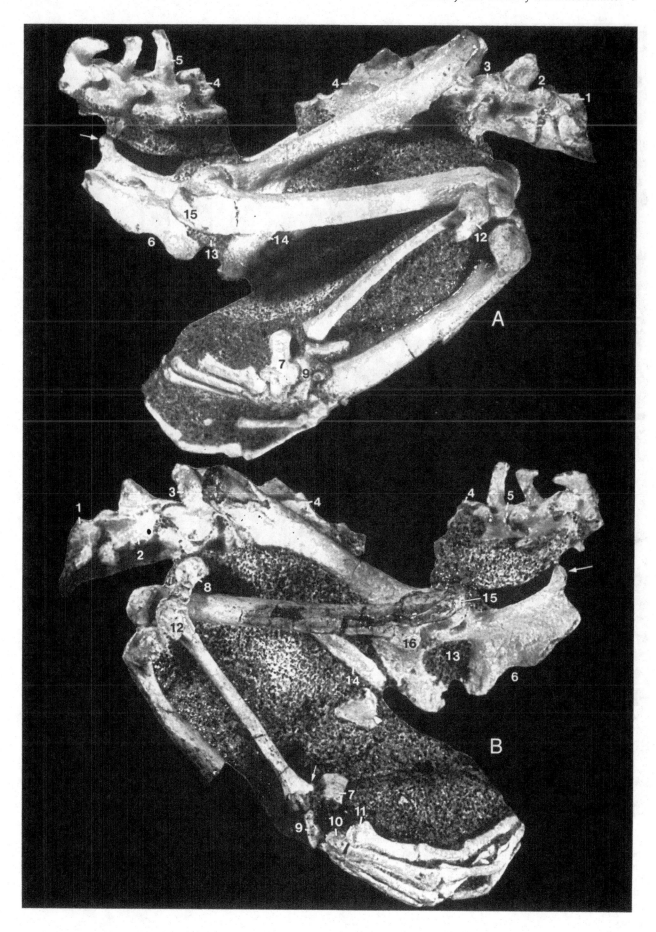

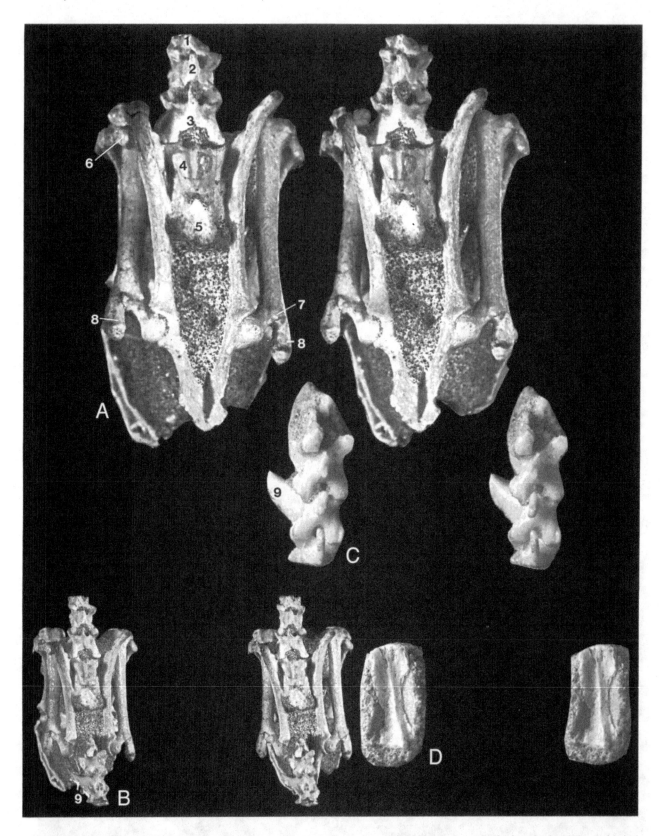

Fig. 3. Kryptobaatar dashzevegi (ZPAL MgM-I/41). □A. Same specimen as in Fig. 2. after removal of fragmentary S4 and Cd1–Cd3, dorsal view. Left ischial tuber seen in Fig. 2 broken off after photographs in Fig. 2 were taken. □B. Same skeleton as in Fig. 2, dorsal view. Note high spinous process in L7 and long transverse process in Cd2. □C. Part of last sacral vertebra and Cd1–Cd3 of skeleton in Fig. 2, after being separated, dorsal view. Note that spinous and transverse processes appear shorter than in B, because of damage caused by preparation. □D. Single middle caudal vertebra found in the same piece of rock as the rest of the skeleton, in ventral view. 1 = L5; 2 = L6; 3 = L7; 4 = S1; 5 = S2; 6 = parafibula; 7 = subtrochanteric tubercle; 8 = greater trochanter; 9 = Cd2, transverse process. A, ×3; B, ×1.5; C, D, ×4; all stereo-pairs.

but now lost, was a roughly parabolic process directed dorso-caudally that overhangs the caudal margin of the ischium (ischial arc) dorsally. The latter is situated at an angle of about 90° with respect to the dorsal margin. The lateral surface of the ischium is concave. To the rear of the obturator foramen there is a prominence, parallel to the caudal margin of the obturator foramen. The inner side of the ischium is slightly concave. The two ischia are firmly fused to form a prominent keel (Figs. 2, 5A) which extends for about 2 mm upwards from the ventral margin. Along the ventral margin of the fused bones there is a distinct, but shallow, indentation

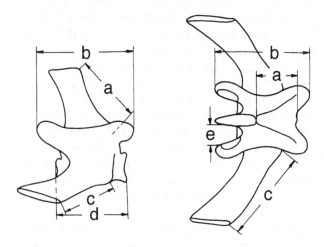

Fig. 4. Diagram showing method of measurement of multituberculate vertebrae, exemplified by lumbar vertebra in left lateral (to the left) and dorsal views. a = length of spinous process; b = distance between prezyga-pophysis and postzygapophysis; c = length of caudal margin of transverse process; d = length of the body; e = distance between prezygapophysis and spinous process.

(postobturator notch), recognized by Krause & Jenkins (1983) as a postobturator foramen. The indentation in *Kryptobaatar* lies within the fused ischiopubic symphysis, and if it is indeed a foramen, it would not open to the pelvic cavity, as characteristic also for other multituberculates (see 'Anatomical comparisons').

The pubis is relatively small, slightly concave in lateral view. Below the obturator foramen it is fused ventrally with its counterpart to form a keel. On the cranioventral margin, dorsal to the junction with the epipubic (marsupial) bone, there is an indentation, 0.7 mm long and 0.3 mm deep. The epipubic bone is 9.3 mm long, boomerang-shaped and slightly asymmetrical. In addition, its margins are fusiform at the extremities and arranged, in lateral view, subparallel to the ilium. In ventral view the epipubic bones meet the symphysis caudally and diverge cranially, which, however, may be due to the state of preservation.

The most characteristic features of the *Kryptobaatar* pelvis are its narrowness, the strong degree of fusion of the ischia and pubes, and the strong fusion with the sacrum. The opposite pubes are directed steeply ventromedially and meet at an angle of about 40°; the ischia meet at 45° and progressively open up to 80° at the ischial arc.

Femur (Figs. 2, 3A, B, 44A). – The femora preserved in acetabula have been pushed medially during sedimentation, and when examined in dorsal view (Fig. 3A, B) they appear parasagittal. The right femur, which is undistorted, is 24.4 mm long. The head is placed on a cylindrical neck that forms an angle of about 50° with the shaft. The head is 2.8 mm wide in dorsal view, its articular surface being greater than a hemisphere. The fovea capitis femoris is not discernible. The greater trochanter (trochanter major) is very prominent, 3.4 mm long, projecting for about 1.2 mm beyond the head on

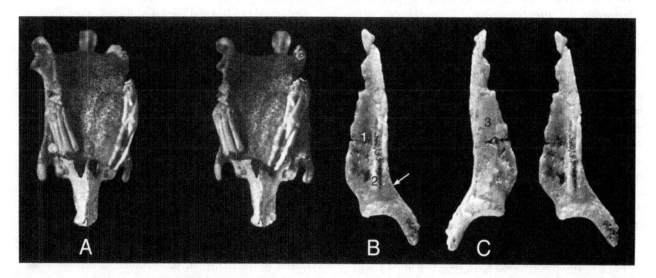

Fig. 5. Kryptobaatar dashzevegi (ZPAL MgM-1/41). □A. Same skeleton as in Figs. 2–4, in ventrocaudal view, showing ventral keel of ischia. □B, C. Partial right scapulocoracoid found in association with same skeleton in laterocaudal and craniomedial views. 1 = ventral part of infraspinous fossa; 2 = broken spine; 3 = subscapular fossa; arrow in B points to incipient supraspinous fossa. A ×1.5; B, C ×4; A and B stereo-pairs.

the left side where it is bent, and for 1.7 mm on the right side. The extremity of the greater trochanter is recurved cranio-medially. Its apex bears a rugose area for insertion of gluteal musculature (Simpson & Elftman 1928). This area is relatively small, roughly crescent-shaped on the dorsal (cranial) aspect, delimited by a sharp crest that overhangs the shaft laterally. On the ventral (posterior) aspect, the rugose area is larger, roughly triangular, tapering distally. Its pointed end merges distally (in lateral view) with a moderately prominent gluteal crest. In dorsal aspect there is a subtrochanteric tubercle (see 'Terminology') and a small, elongated fossa lateral to it.

The lesser trochanter (trochanter minor), as seen in latero-ventral view, is very prominent, 1.9 mm long and 1.9 mm wide, arising from the mid–width of the ventral wall of the shaft (number 16 in Fig. 2B, see also the femur of *Nemegtbaatar*, number 3 in Fig. 16A, C, for comparison). It forms a plate-like, bent process, strongly protruding ventrally, convex lateroproximally and concave mediodistally; in caudal view it has a hook-like profile. The lesser trochanter is not constricted at the base, and the neck, characteristic of ?*Eucosmodon* (Granger & Simpson 1929), is not present.

The trochanteric (digital) fossa (fossa trochanterica) is small, roughly triangular and poorly preserved; it prolongs as a narrow groove onto the greater trochanter. We designate the fissure-like fossa situated lateral to the lesser trochanter, the *post-trochanteric fossa* (referred to by Simpson & Elftman 1928 and Krause & Jenkins 1983 as a caudal part of the divided trochanteric fossa; see 'Anatomical comparisons'). The post-trochanteric fossa appears to be shallower in *Kryptobaatar* than in other genera, which may be due at least in part to the state of preservation. The shaft is elliptical in cross-section, dorsoventrally compressed, more convex dorsally than ventrally, 2.8 mm wide and 1.9 mm deep in the middle of its length.

The boundary between the shaft and the distal epiphysis is recognizable, the latter being twisted slightly laterally with respect to the proximal epiphysis. The width of the distal epiphysis in dorsal view is 5.4 mm. The trochlea is shallow, and the low trochlear ridges are arranged subparallel (rather than obliquely) with respect to the shaft. There is a wide intercondyloid fossa (fossa intercondylaris). The lateral condyle (condylus lateralis) is in dorsal view larger than the medial one (condylus medialis) and protrudes laterally over the shaft for 1.4 mm. In distal view the medial condyle is wider than the lateral one; the width of the whole epiphysis in this view is 4.9 mm, the width of the lateral condyle is 1.9 mm and of the medial condyle 1.9 mm (Figs. 2 and 3A, but see also the femur of *Nemegtbaatar* Fig. 17B–E, for comparison). The femur articulates only with the tibia; there is no femoro-fibular contact.

Tibia (Figs. 2, 6C, 44A). – The right tibia is 18.8 mm long; its proximal epiphysis is very large, asymmetrical and 5.1 mm wide in cranial view. The proximal articular surface is di-

vided by a weak protuberance, directed mediocaudally to craniolaterally, into two facets. The small medial facet, for articulation with the medial femoral condyle, is arranged parallel to the above mentioned protuberance. The larger lateral facet, for the lateral femoral condyle, is transversely elongated and has an irregular proximal surface with a distinct, transverse groove in the middle. Laterally it extends into a prominent hook-like process, which is triangular in lateral view and overhangs the shaft laterally. It bears a facet for the fibula on its caudal surface. The process is concave from below, and in ventral view a concave bridge of bone spans the distance between the process and the proximal end of the tibia. The proximal articular surface extends caudally well beyond the shaft. At the confluence of the shaft and the proximal articular area there is thus a broad, deeply concave fossa. A ridge of bone separates this fossa from one adjacent to it and beneath the aforementioned hook-like process. The cranial (anterior) crest (margo cranialis, crista tibiae), because of the caudal concavity of the proximal part of the shaft, is comparatively sharp proximally, but it is poorly exposed as both tibiae are partly embedded in the matrix. The cranial (anterior) tibial tuberosity (tuberositas tibiae) is hardly recognizable, although the cranial surface of the tibia is convex. The shaft is compressed craniocaudally at the middle of its length, its craniocaudal diameter is 2.1 mm, the transverse diameter 1.5 mm; in lateral view it forms a gentle convex arch.

The distal part of the right tibia is preserved; the medial malleolus has been broken off and is preserved on the dorsal surface of the astragalus (number 1 in Fig. 6A–C). A piece of the flat bone preserved between the distal parts of the tibia and fibula (Fig. 2A) is possibly a displaced fragment of the right dentary. There is a transverse seam across the tibia, separating the distal articular surface from the shaft, which might be at least in part a remnant of the boundary with the epiphysis.

The lateral condyle and medial malleolus of the tibia are less obvious than on the tibia of North American ?*Eucosmodon* figured by Krause & Jenkins (1983, Fig. 22; see also Fig. 56E, F herein). The lateral condyle is large but not well defined (see also 'Anatomical comparisons').

Fibula (Figs. 2, 6C, 44A). – Both fibulae are preserved almost in anatomical position proximally, but slightly displaced distally. The length of the left one, which preserves the distal caudal tuberosity, is 17.2 mm. The fibula is situated lateral to the tibia proximally, and more caudally in the distal part. The head of the left fibula is 2.4 mm wide in lateral view. It bears on its lateral side a prominent, triangular, hook-like process (partly broken), that extends distally. Proximally on the cranial wall of the process there is a flat facet articulating with a large facet on the laterocaudal wall of the tibial head. Posterior to the triangular process there is another small triangular process on the distal border of the head, directed distally. In lateral view, the large triangular processes of the

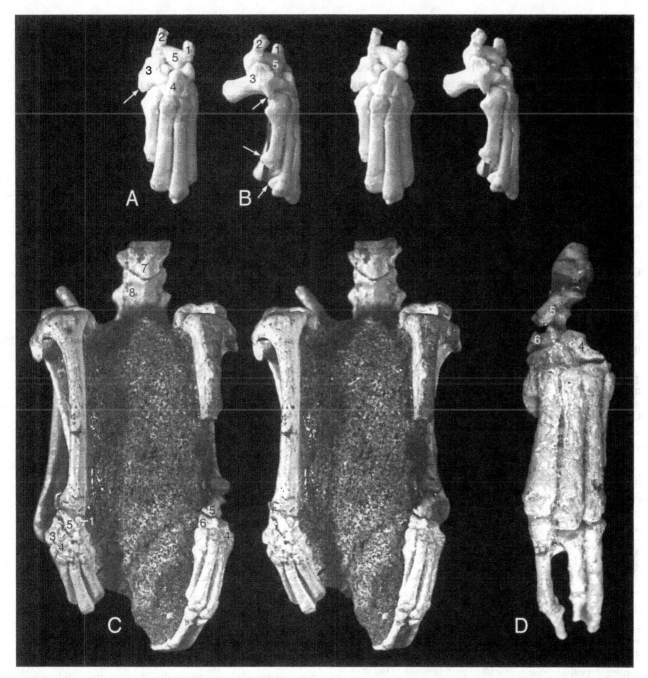

Fig. 6. □*Kryptobaatar dashzevegi* (ZPAL MgM-I/41). □A, B. Incomplete right pes isolated from skeleton in C, in dorsal and lateral views. □C. Ventral view of skeleton in Fig. 2, before separation of right pes. □D. Left pes of skeleton in C, with metatarsals placed horizontally. 1 = part of medial malleolus of tibia; 2 = part of lateral malleolus of fibula; 3 = calcaneum; 4 = cuboid; 5 = astragalus; 6 = navicular; 7 = L5; 8 = L6. Arrow in A and upper arrow in B = peroneal tubercle; two lower arrows in B = sesamoid bones. A, B ×4; C ×3; D ×6; A–C, Stereo-pairs.

tibia and fibula are aligned. On the caudal wall of the fibular head there is a large facet for articulation with the parafibula (preserved on the left side, see below). Extending distally along the middle of the lateral wall of the shaft, beginning below the large triangular process, there is a rounded ridge with proximal concavities on both sides. There are sharp short ridges bounding these concavities, one on the cranial,

the other on the caudal wall of the shaft. At the middle of its length the shaft becomes thinner and roughly parabolic in cross section, compressed craniocaudally; it is 0.9 mm wide in lateral view. Distally the diameter of the shaft increases.

The distal epiphysis is almost completely preserved on the left side, and the lateral malleolus has been preserved (the arrow in Fig. 2B). The middle part of the epiphysis is developed as a rounded tuberosity, convex distally; this is separated from the shaft by a transverse line that may correspond to the boundary of the shaft with the epiphysis. On the right side a part of the lateral malleolus has been broken off and is preserved laterally on the dorsal surface of the astragalus (number 2 in Fig. 6A, B).

Parafibula. – The parafibula has been preserved only on the left side of ZPAL MgM-I/41 (Figs. 2B, 3A, B, 44A). It is displaced laterally and arranged at an angle of about 125° with respect to the fibular shaft. The length of the ossicle is 3.2 mm. It consists of a relatively narrow neck, 1.2 mm wide, and a large, rounded muscular process. Along the middle of the neck, there extend irregular grooves, roughly parallel to its margin. The muscular process is concave at the contact with the neck and is inflated, but irregular, at the end; the inflated part is 2.2 mm wide. The articular facet, which adheres to the craniolateral surface of the fibular head, cannot be examined. This parafibula differs from that of *Ptilodus kummae* (Krause & Jenkins 1983, Fig. 23F) in having a more rounded and inflated muscular process, rather than one obliquely cut at the end.

Tarsus and pes
Figs. 2, 3A, B, 5A, 6, 7, 54, 59A

On the right side all seven tarsal bones and five metatarsals are preserved. The tarsal bones adhere tightly to each other and are preserved in the original position, except for the calcaneum that has been shifted transversely and the astragalus which has been shifted laterally and in a plantar direction. As a consequence, the mediodistal rounded process (referred to by Granger & Simpson 1929 and Krause & Jenkins 1983 as the astragalar head; see 'Anatomical comparisons') has been exposed. On the left side (Figs. 2B, 3A, B, 5A, 6C–D) an almost complete tarsus, five metatarsals and all the phalanges of D II–V, are preserved. Of the calcaneum only the tuber has been preserved; all other bones are in place, but the astragalus has been slightly displaced longitudinally and in a plantar direction, showing the mediodistal triangular process. The astragalus contacts the navicular with its shorter (mediodistal) margin. On both sides the tarsus and pes are strongly convex dorsally, possibly, at least in part, because of the state of preservation. Therefore in Figs. 6 and 7 only a part of the tarsus is visible in each view (see reconstruction in Fig. 54). On the left side, the tarsus and pes have been left as originally preserved and can be examined only in dorsal aspect. On the right side, after the photographs in Figs. 2 and 6C were taken, the tarsus and partial pes were separated from the rest of the specimen, but as the tarsal bones could not be separated, they can hardly be seen in plantar aspect (Figs. 6A, B and 7A–D). The description of bones that follows, unless stated otherwise, is of the dorsal (cranial) aspect. In Fig. 54 we give the reconstruction of the pes of *Kryptobaatar* based on ZPAL MgM-I/41 and on the tarsus of *Chulsanbaatar* ZPAL MgM-I/99b (Fig. 25). See 'Functional anatomy' for discussion of the function of the pes of *Kryptobaatar*.

Calcaneum. – The calcaneum is about 4.6 mm long (right pes, Figs. 2A, 6A–C, 7A–D, 54) and convex dorsally along its longitudinal axis. The robust tuber calcanei is laterally compressed. Its proximal end is slightly enlarged, irregular, and bears several small, poorly pronounced tuberosities. Extending obliquely along the dorsal surface of the tuber calcanei is a weak ridge that ends at the dorsal margin of the peroneal tubercle. The peroneal tubercle, as preserved, forms a prominent, sharply pointed triangular process, strongly protruding obliquely laterodistally beyond the lateral margin of the bone (arrow in Fig. 6A and upper arrow in Fig. 6B). There is a weak ridge extending transversely across the peroneal tubercle. The poor state of preservation of the bone does not allow us to state with any certainty whether the peroneal tubercle has been broken. As in all known multituberculate calcanea (Granger & Simpson 1929; Deischl 1964; Krause & Jenkins 1983; Szalay 1993; this paper Figs. 25, 55F–I, 56H–J) this tubercle is rounded at the end, we presume that it has been broken in *Kryptobaatar* and that its distal part is missing. This is why the reconstruction of the peroneal tubercle in Fig. 54 is based on better preserved North American multituberculate calcanea. However, the peroneal groove of *Kryptobaatar* differs from that in ?*Eucosmodon* in being wider, and in consequence the peroneal tubercle is shifted more laterally. In this respect, the calcaneum of *Kryptobaatar* is reminiscent of the unidentified multituberculate calcaneum from the Hell Creek Formation (Fig. 55E–G) and of a ptilodontid calcaneum from the Frenchman Formation (Szalay 1993, Fig. 9.8). Although the calcaneum of *Kryptobaatar* has been shifted laterally, it is evident that its distal margin (medial to the peroneal groove) articulated with the small facet on the dorsal side of the proximal end of Mt V. The medial part of the distal margin of the calcaneum possibly articulated with a wide indentation on the lateral margin of the cuboid, proximal to the facet for articulation with Mt V. The sustentaculum was probably small, but the facet cannot be examined. The astragalocalcaneal facet (proximal facet for the astragalus, facies articularis talaris) is very large and crescent-shaped but less prominent than in ?*Eucosmodon* and in various calcanea figured by Krause & Jenkins (1983, Fig. 26). Distal to it, there is a smaller cuboid facet, which was possibly situated more distally than in the unidentified North American calcanea. Between the astragalocalcaneal facet and the peroneal tubercle there is a wide, shallow concavity, delimited distally by a faint ridge. Distal to this ridge the surface of the bone bends downwards. The plantar side of the *Kryptobaatar* calcaneum cannot be examined.

Astragalus (talus). – Because of the oblique position of the astragalus in respect to the calcaneum (see reconstruction in Fig. 54), its margins are not oriented distally, proximally, laterally and medially, but rather medioproximally, laterodistally, etc. The astragalus can be examined in both pedes in dorsal (cranial) views, in the left pes also in a distoplantar view and in the right pes in proximal and distal views (Figs. 2, 6, 7). In the right pes, on the dorsal side, the medial malleolus

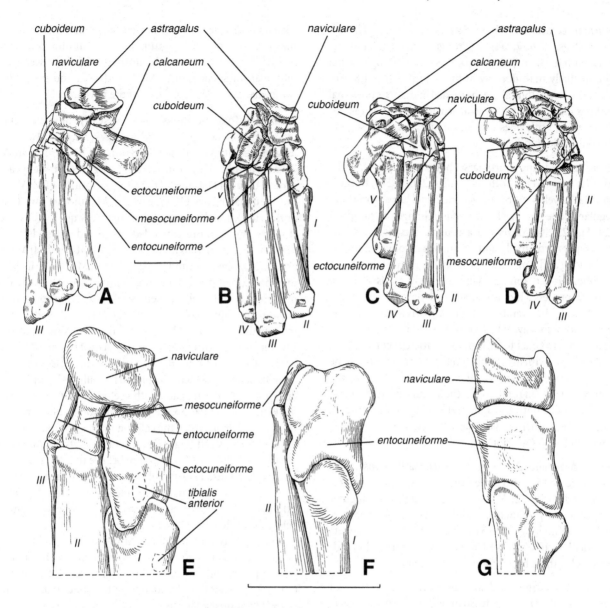

Fig. 7. Kryptobaatar dashzevegi (ZPAL MgM-I/41), Isolated elements of skeleton in Fig. 2. □A–D. Camera lucida drawings of right tarsus and metatarsals. Fragments of distal ends of tibia and fibula preserved on dorsal surface of astragalus, seen in Fig. 6A, B, are omitted. Roman numerals denote metatarsals. □A. Medial view. □B. Dorsal view. □C. Dorsolateral view. □D. Lateral view. □E–G. Details of entocuneiform–Mt I joint, dorsal (slightly medial), mediodorsal and medial views. In F the navicular is omitted. Scale bars 2 mm.

of the tibia and part of the lateral malleolus of the fibula have been preserved (Fig. 6A–C). In dorsal aspect the astragalus forms an irregular rectangle, arranged roughly laterodistally to medioproximally in the reconstructed pes. The mediodistal margin of the rectangular part is thickened. In laterodistal view the rectangular part sends a rounded process ('astragalar head'), situated more medially than laterally. Along the cranial margin of this process there is a saddleshaped sulcus, which prolongs medially below the mediodistal thickened end of the rectangular part and articulates with the navicular. The details of the 'astragalar head' are

better seen in isolated astragali of an unidentified multituberculate from the Late Cretaceous of North America (Fig. 55B–D) and of Paleocene *?Eucosmodon* sp. (Fig. 56C, D). In proximal aspect the astragalus of *Kryptobaatar* is roughly triangular, with a rounded tip pointing in a plantar direction. The astragalocalcaneal and sustentacular facets are not exposed on either side. The dorsal surface is relatively smooth, especially in comparison with *?Eucosmodon* sp. (Fig. 56C, D), but both lateral and medial malleolar facets are recognizable, the lateral large and gently rounded, the medial much smaller and more prominent.

Navicular (central tarsal bone, os tarsi centrale). – The navicular (Figs. 2, 6, 7, 54) is a roughly rectangular bone, preserved almost in place on both sides, although facing more medially than dorsally in the right pes. It is 1.8 mm wide and 1.1. mm long along the medial margin. Its proximal margin is incurved medially to articulate with the astragalus (displaced on both sides); the distal margin is only slightly incurved in the middle.

Cuboid (cuboideum, fourth tarsal bone, os tarsale 4). – The cuboid is exposed on both sides (Figs. 2, 6, 7, 54), but it is particularly well seen on the left side and has a somewhat irregular shape (as in ?*Eucosmodon*, Granger & Simpson 1929, Fig. 23). It consists of a roughly rectangular body, which tapers proximally and is somewhat convex dorsally, 1.8 mm long and 1 mm wide proximally. The distal width of the cuboid is 1.6 mm. Distally the body sends a lateral transverse process which articulates with the medioproximal facet of Mt V. The lateral side (exposed in the left pes) distally forms a deep concavity to receive the distal margin of the calcaneum. Proximal to this concavity there is convex facet for articulation with the cuboid facet of the calcaneum. In the right pes the distal margin of the calcaneum (although the calcaneum has been rotated), fits the lateral concave facet on the cuboid. On both sides, the remaining tarsal bones fit tightly to each other, and the facets for ectocuneiform, navicular, and Mt IV are not seen.

Ectocuneiform (lateral cuneiform, third tarsal bone, os tarsale 3). – The ectocuneiform is preserved in anatomical position on both sides (Figs. 2, 6, 7, 54). It is a roughly rectangular bone, rounded proximally, slightly wider proximally than distally, 1.3 mm long and 0.5 mm wide distally. It snugly fits the navicular, cuboid, Mt III and mesocuneiform, although the facets for these bones cannot be recognized.

Mesocuneiform (intermediate cuneiform, second tarsal bone, os tarsale 2). – The mesocuneiform is roughly rectangular, 0.8 mm long and 0.8 mm wide in the middle, slightly wider distally than proximally, rounded at the corners (Figs. 2, 6, 7, 54). It fits tightly between the navicular, ectocuneiform, entocuneiform and Mt V although its facets are not exposed.

Entocuneiform (medial cuneiform, first tarsal bone, os tarsale 1). – The entocuneiform is the longest of the three cuneiforms. It notably protrudes over the distal margin of other cuneiforms and is laterally compressed (Figs. 2, 6, 7, 54). The distal surface is saddle-shaped in the dorsoplantar direction; in the middle of the distal margins of the dorsal and plantar sides there are processes (the plantar longer) with a concavity between them. As a result, in medial view the distal margin is strongly concave (Fig. 7A, B, E–G). These processes embrace the saddle-shaped proximal margin of Mt I, on which there are lateral and medial processes. The proximal margin is slightly concave. In dorsal view the entocuneiform, as preserved, is oriented obliquely dorsomedially and is narrower than in medial view. In medial view the plantar margin is 1.9 mm long, the dorsal margin is 1.4 mm, and the proximal width is 1.5 mm. The entocuneiform fits tightly between the navicular, mesocuneiform, Mt II and Mt I and protrudes somewhat over the medial side of the navicular.

Metatarsals (ossa metatarsalia I–V). – Nearly complete metatarsals are preserved on both sides (Figs. 2, 5A, 6, 7, 54). Their lengths are: Mt I 5.4 mm, Mt II 6.8 mm, Mt III 7.4 mm, Mt IV 7.3 mm and Mt V 5.4 mm. In all the figures Mt V appears shorter than Mt I because of the oblique position of Mt V. Of the five metatarsals Mt V is the widest, Mt I the narrowest (in dorsal view), and the three medial are of subequal widths. Mt II – Mt IV are more expanded at their distal ends than proximally, but there is no distal groove. Mt I and Mt V are expanded at the distal and proximal ends. Mt I has a distinct medioproximal process and apparently also a lateroproximal process, not well exposed. Mt V has a lateroproximal prominence, but smaller than in Mt I. Mt V is shifted proximally with respect to Mt IV, and the medioproximal facet on Mt V articulates with the laterodistal facet of the cuboid. The proximal tip of Mt V fits the medial part of the distal margin of the calcaneum (which is displaced laterally and in a plantar direction). Mt V articulates with the peroneal tubercle and fits with its proximal tip the peroneal groove. Similarly, in *Chulsanbaatar*, in which the calcaneum has been somewhat differently displaced (Fig. 25), Mt V fits with its proximal tip the peroneal groove. We therefore believe that Mt V articulated with the distal margin of the calcaneum, medial to the peroneal groove in both taxa (see reconstruction in Fig. 54). Mt I and Mt V have been preserved in both pedes slightly below the middle ones (in dorsal view).

The joint between the entocuneiform and Mt I is almost of hinge type and allowed extensive movements in the dorsoplantar plane, with the possibility of only very little abduction. We have not found evidence for the opposability of Mt I (see 'Functional anatomy').

Sesamoid bones (Figs. 6B, 54B–D). – On the right side of ZPAL MgM-I/41 the sesamoid bones have been preserved on the plantar side at the distal ends of all metatarsals except Mt I.

Phalanges (Figs. 2, 3A, B, 5A, 6C, D, 54). – The digits have been preserved in the left pes, but D II is missing, only the metatarsal having been preserved. In D I and D III the ungual phalanges are missing, while D IV and D V are complete. In D II, Ph 1 is slightly damaged, but its length can be estimated, as Ph 2 and Ph 3 have been preserved in place. All the phalanges are more expanded proximally than distally and taper distally. In D II, Ph 1 and Ph 2 are 4.5 and 3.8 mm long, respectively. In D III, Ph 1 and Ph 2 are 5.1 and 2.8 mm respectively. In D IV, Ph 1, Ph 2 and Ph 3 are 4.1, 2.7 and 1.3 mm long. In D V, Ph 1, Ph 2 and Ph 3 are 3.1, 2.5 and 1.3 mm long. The ungual phalanges are pointed and roughly rounded in cross section.

Genus *Nemegtbaatar* Kielan-Jaworowska, 1974

Nemegtbaatar gobiensis Kielan-Jaworowska, 1974

Figs. 8–13, 14A, 15, 16, 17B–I, 28C, 30B, 34A, 35, 36B, 37A, 39–43, 45, 47, 50, 61

Material. – Red beds of Khermeen Tsav, Khermeen Tsav II, ZPAL MgM-I/81: skull associated with dentaries, fragment of C1, C2–C4, ?L2–L7, S1, damaged S2, S3 and S4, fragments of ribs, ventral part of the right scapulocoracoid, proximal part of the right humerus and, separately, its damaged distal part, proximal part of the left ulna with broken olecranon associated with a proximal part of the radius, distal part of the right radius associated with the ?scaphoideum and ?lunatum, two isolated carpal bones (?triquetrum and displaced ?pisiform, ?trapezoideum or ?praepollex), an incomplete pelvis, the proximal part of the right femur, and the middle part of the shaft of the left femur. ZPAL MgM-I/82: skull of a juvenile individual, both dentaries, a damaged atlas, C2–C7, T1–T4, rib fragments. ZPAL MgM-I/110: the left femur in two pieces, the distal part, in anatomical articulation with the tibia and fibula and a fragment of a ?parafibula preserved in the knee joint. ZPAL MgM-I/110 is smaller than ZPAL MgM-I/81 but possibly corresponds to the size of a juvenile ZPAL MgM-I/82. The assignment of ZPAL MgM-I/110 to *N. gobiensis* is tentative, based on size and on the fact that *Nemegtbaatar* is the only multituberculate of such a size in the Barun Goyot Formation.

Skull

Figs. 8A, D

The skull of *Nemegtbaatar* was described by Kielan-Jaworowska (1974), Kielan-Jaworowska *et al.* (1986), and Hurum (1992, 1994). In ZPAL MgM-I/81 its estimated width is 31 mm, the estimated length is 40 mm, which corresponds to 1.15 of the estimated length of the pelvis.

Axial skeleton

Cervical vertebrae and cervical ribs
Figs. 8B–F, 9, 10

In ZPAL MgM-I/82 C2–T4 are preserved together, and the whole set is bent dorsally. In ZPAL MgM-I/81, C2, C3 and C4 are in succession but are separated from each other. Between C2 and C3 a short piece of bone separating the two bodies is preserved in the position of an intervertebral disc.

Atlas (Fig. 8B–D). – In ZPAL MgM-I/81 a fragment of a right side of the atlas, with cranial and caudal articular foveae and a transverse process, is preserved (Fig. 8B–C). The cranial articular fovea is roughly oval, concave, narrowing dorsally; its longer diameter measures about 3.3 mm. The transverse

process is peg-like, about 2.2 mm long, and is rounded at the end. The caudal articular fovea is roughly oval, slightly narrowing ventrally, almost flat; its longer diameter is 2.0 mm. There is a small foramen (about 0.2 mm in diameter, number 9 in Fig. 8B, C) situated dorsal to the caudal articular fovea and a corresponding foramen situated laterodorsally to the cranial articular fovea. Both are apparently too small to transmit the arteria vertebralis and possibly are vascular foramina (see 'Anatomical comparisons'). In ZPAL MgM-I/82 (Fig. 8D; see also Kielan-Jaworowska *et al.* 1986, Fig. 18) an incomplete atlas, lacking the ventral arch and the transverse processes, is broken into several pieces and was not separated from the skull.

Axis. – In ZPAL MgM-I/81 (Fig. 10) the axis (without the dens) is 2.8 mm long and in ZPAL MgM-I/82 (Fig. 8F, 9) 1.3 mm long. The base of the dens is in ZPAL MgM-I/82 about 1.4 mm wide and 0.9 mm deep. The body is strongly compressed dorsoventrally.

The cranial right and left articular surfaces are confluent with each other and form, in ventral view, a crescent-shaped area that strongly protrudes ventrally; in ZPAL MgM-I/82 the ventral part of the body to the rear of this area is partly damaged. The ventral crest is prominent and widens both cranially and caudally. The depressions lateral to the crest are deep, especially immediately behind the cranial articular surfaces, and are pierced by numerous nutrient foramina. The transverse processes are broken off in both specimens, but their bases are well preserved in ZPAL MgM-I/81. As in all the cervicals, the transverse processes arise by two roots: a ventral one from the body, visible in ventral view as a prominent oblique crest, and a dorsal one from the arch. The transverse canal is present; it is open ventrally on both sides of ZPAL MgM-I/81 because of the damage; a fragment of the cervical rib (see below) has been preserved on the right side. The postzygapophysis is large, directed cranioventrally. The dorsal side of the body is flat, perforated by one or two pairs of nutrient foramina. The arch is preserved in ZPAL MgM-I/81, the pedicles are moderately high, and the laminae arise obliquely dorsomedially above the posterior part of the body and the postzygapophysis. The laminae are more stout at the base than upwards, roughly parallel–sided, about 3.9 mm long at their longitudinal midpoint. At the point where right and left laminae meet, the length of the arch increases insignificantly. At the top there is a broken surface, forming the base of the spinous process, which, judging from the size of the arch and preserved broken surface, appears to have been relatively small. The vertebral foramen as preserved in ZPAL MgM-I/82 was relatively high, distinctly narrowing dorsally.

C3–C6 (Figs. 8E, 9, 10). – The body of C3 is slightly less compressed dorsoventrally than in C2; the degree of dorsoventral compression of the successive vertebrae gradually decreases caudally along the vertebral column. The body of C3 (in ventral view) is 2.2 mm long in ZPAL MgM-I/81 (Fig. 10) and 1.7 mm long in ZPAL MgM-I/82 (Fig. 9), C4 and C5

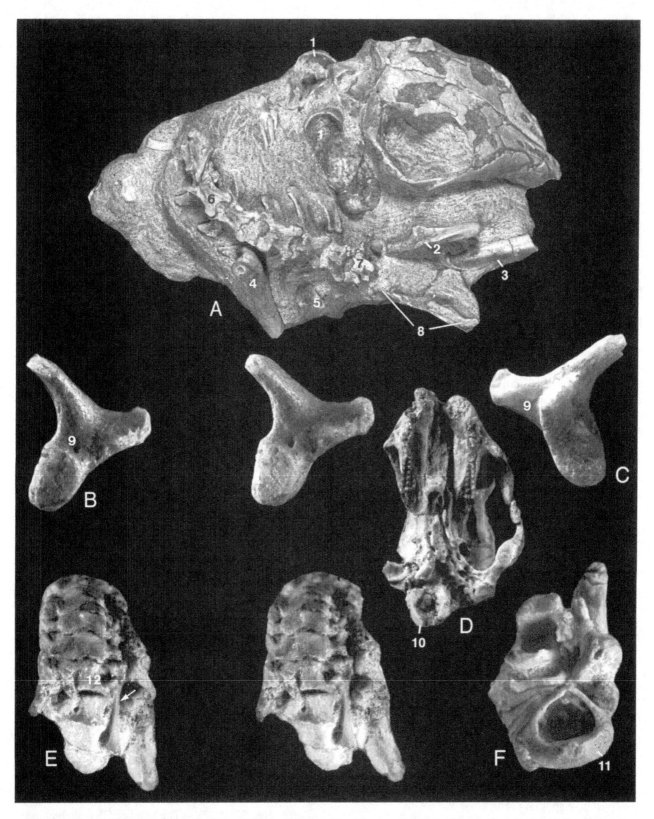

Fig. 8. Nemegtbaatar gobiensis. □A–C. ZPAL MgM-I/81, Red beds of Khermeen Tsav, Khermeen Tsav II, Gobi Desert, Mongolia. □A. Skull and incomplete skeleton as found, after partial preparation; fragments of damaged ribs not marked by numerals are also seen. □B, C. Right side of atlas of individual in A, caudal and cranial views. □D–F. ZPAL MgM-I/82, horizon and locality as in A. □D. Skull and atlas of juvenile individual in ventral view. □E, F. Cervical and thoracic vertebrae of individual in D, ventrocaudal and cranial views. 1 = dentaries (only posterior border of left one is seen); 2 = anterior part of right ilium; 3 = proximal part of right femur; 4 = distal part of left femur, epiphysis of which was lost during preparation; 5 = anterior part of left ilium seen in cross-section; 6 = L3; 7 = S1; 8 = damaged S2–S4; 9 = ?vascular foramen; 10 = atlas; 11 = axis with broken dens; 12 = T1; arrow in E points to first left thoracic rib. A ×1.5; B, C ×8; D ×2; E, F ×4; B and E stereo-pairs.

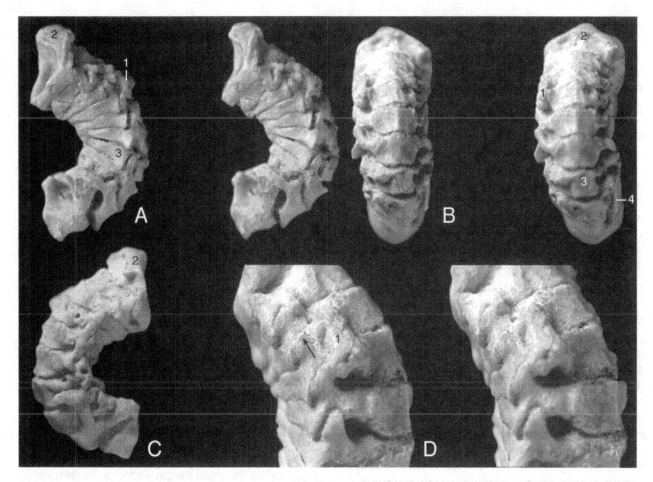

Fig. 9. Nemegtbaatar gobiensis (ZPAL MgM-I/82). □A–C. Cervical and first thoracic vertebrae C2–T4 of individual in Fig. 8D–F, left lateral, ventral and right lateral views. Matrix with bone fragments seen on left side of specimen (right side in photographs) in Fig. 8E and F has been removed. □D. Middle part of same specimen in oblique lateroventral view. 1 = fourth cervical rib; 2 = axis; 3 = first thoracic vertebra; 4 = first left thoracic rib; arrow in D = groove for arteria vertebralis (transverse canal). A–C ×4; D ×8; all except C are stereo-pairs.

in ZPAL MgM-I/82 are 1.7 mm long each, C6 1.9 mm. The width of the bodies in ZPAL MgM-I/82 (Fig. 9) are: C3 3.8 mm, C4 3.8 mm, C5 4.2 mm and C6 3.2 mm. The bodies of C3–C4 are roughly rectangular, those of C5–C6 are gradually more trapezoidal, their cranial diameters becoming wider in relation to the caudal diameters. The ventral crest on C3 is less prominent than on C2, but on successive vertebrae it becomes more prominent up to C6. The depressions lateral to the crest are shallower on C3–C6 than on C2. The ventral branch of the transverse process on all the cervicals is prominent, but true inferior (ventral) lamellae are not developed (Howell 1926; see also Kielan-Jaworowska 1977). On C3 the ventral branch extends obliquely along almost the whole length of the body; on the successive vertebrae it is gradually shorter, and on C4–C6 it is confined to the first two-thirds of the body length. The base of the dorsal branch of the transverse process is shorter than the ventral one and arises on all the vertebrae below the most caudal part of the prezygapophysis. The transverse processes are broken off at various levels on all the vertebrae, except for the right side of C4 in ZPAL Mg-MI/82, where the cervical rib (see below) has been

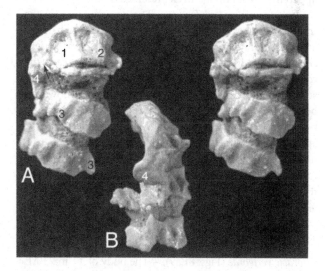

Fig. 10. Nemegtbaatar gobiensis (ZPAL MgM-I/81). □A. Cervical vertebrae C2–C4 of individual in Fig. 8A, ventral view. □B. Same specimen in left lateral view. 1 = depression lateral to ventral crest; 2 = ventral root of transverse process of axis; 3 = ventral root of transverse processes of C3 and C4; 4 = fragment of broken cervical rib; arrow in A = groove for arteria vertebralis (transverse canal). Both ×4; A stereo-pair.

preserved. Although the transverse processes are broken, the groove for the arteria vertebralis and its accompanying vein (part of the transverse canal) is clearly seen between the bases of the ventral and dorsal branches of the transverse processes. The pre- and postzygapophyses are well preserved on both sides of almost all the cervicals in ZPAL MgM-I/82 and less completely in ZPAL MgM-I/81. They form large oval surfaces arranged obliquely cranioventrally. The relatively narrow laminae arise dorsomedially and slightly above the postzygapophyses. In C3 and C4 the arches are of almost the same length all along their height, whereas in C5–C6 their diameters decrease dorsally. On C6 there is a tubercle at the point where the left and right laminae meet. Traces of the tubercles are also present on the anterior cervicals. From C2 to C7 the laminae become gradually more horizontally arranged and the vertebral foramen becomes lower.

C7. – The body of C7 in ZPAL MgM-I/82 (Fig. 9) is 2.0 mm long and 2.9 mm wide, trapezoidal and generally similar to that of C6. The broken-off ventral branch of the transverse process is preserved on both sides and extends obliquely from and along little more than a half of the body length. The dorsal branch is shorter than the ventral one, broken on the right side, but preserved on the left side as a triangular process directed dorsocaudally, overlapping ventrally the transverse process of T1. There is a distinct longitudinal groove between the dorsal and ventral branches of the transverse process, although it is shallower and less distinct than on the preceding vertebrae. We therefore conclude that the transverse process was perforated by the transverse foramen. The caudal costal fovea is not recognizable on C7 with any certainty.

Cervical ribs. – In ZPAL MgM-I/81 a broken part of an apparent cervical rib has been preserved on the right side of the axis (number 4 in Fig. 10). The dorsal branch of the transverse process has been preserved on C3, and both dorsal and ventral branches are preserved on C4 and C5 on the right side in ZPAL MgM-I/82 (Fig. 9A, B, D). In all these cases the transverse processes end with distinct, rounded articular surfaces for the cervical ribs. The damaged rib has, however, been preserved only between C4 and C5 and apparently belongs to C4; because of its poor state of preservation (it appears to be pointed cranially rather than caudally) it cannot be excluded that this is the rib of C5 which has been turned around. The preserved part of the cervical rib shows that it was possibly relatively large, roughly oval in lateral view.

THORACIC VERTEBRAE AND RIBS

Thoracic vertebrae (Figs. 8E, 9). – The body of T1 in ZPAL MgM-I/82, is in ventral view, similar in outline to that of C7, 2.2 mm long, roughly rectangular, 3.3 mm wide. The middle part of the body protrudes ventrally. The ventral crest is missing. At the craniocaudal corners of the body there are, on both sides, what appear to be 'tubercles' which we recognize

as broken off heads of the first ribs, cemented to the cranial costal foveae. If so, the cranial costal foveae were large, facing mostly ventrally. The caudal costal foveae are hardly discernible on either side, possibly very small. The transverse process is single-pronged, imperforate and arises opposite the most cranial part of the vertebral body (ventral view). On the left side, the costal fovea of the transverse process is preserved in articulation with the first rib, facing ventrocaudally. The prezygapophysis is broken off on the right side and damaged on the left. The postzygapophysis is very large, directed obliquely cranioventrally, as on the cervical vertebrae. The lamina arises obliquely dorsomedially above the postzygapophysis; its length decreases upwards. There is a tubercle at the top of the arch. The arch is slightly lower than on C7 and the anterior cervicals.

T2 is similar to T1, its body being 2.2 mm long and 3.6 mm wide. The arch is lower than in T1, the laminae being arranged more horizontally than vertically, with a small tubercle (spinous process) on the top.

The bodies of the consecutive thoracic vertebrae T3 and T4 become gradually more compressed laterally and longer; T3 is 2.3 mm long and 3.6 mm wide, T4 (damaged) is in caudal view roughly triangular, tapering ventrally, with small caudal costal foveae at the dorsolateral corners. On T3 the caudal costal foveae are partly seen, and the cranial costal foveae on both vertebrae are poorly recognizable. The arch, preserved only on T3, is low, the laminae being arranged almost horizontally. There is a dorsal tubercle and minute 'processes' on the caudal margin. Otherwise the two vertebrae do not differ from those of T1 and T2.

Thoracic ribs (Figs. 8E, 9B, C). – The proximal part of the first left rib in ZPAL MgM-I/82 (the arrow in Fig. 8E) is arranged longitudinally along the lateral side of the bodies of T1 and T2. The head is cemented to the cranial costal fovea of T1, and the neck (displaced) articulates with it, while the displaced tubercle articulates with the costal fovea of the transverse process of T1. In addition to the heads of the right first and second ribs cemented to the respective costal foveae in the same specimen, numerous broken rib fragments are preserved in ZPAL MgM-I/81.

LUMBAR VERTEBRAE
Figs. 8A, 11, 34A, 35, 36B, 39B, 40B

We tentatively accept that there were seven lumbar vertebrae in *Nemegtbaatar*, the first one not being preserved. The arches, when seen in cranial and caudal views, form a low rectangle. The pedicles are strongly concave laterally but are vertically expanded. On the boundary between the pedicle and lamina there is a prominent intermediate crest, developed as an edge, concave in the middle, that extends between the lateral margins of the pre- and postzygapophyses. On the concave surface of the pedicle, below the intermediate crest there is a weak longitudinal ridge, convex dorsally, that is well preserved only on the right side of L4 and L5.

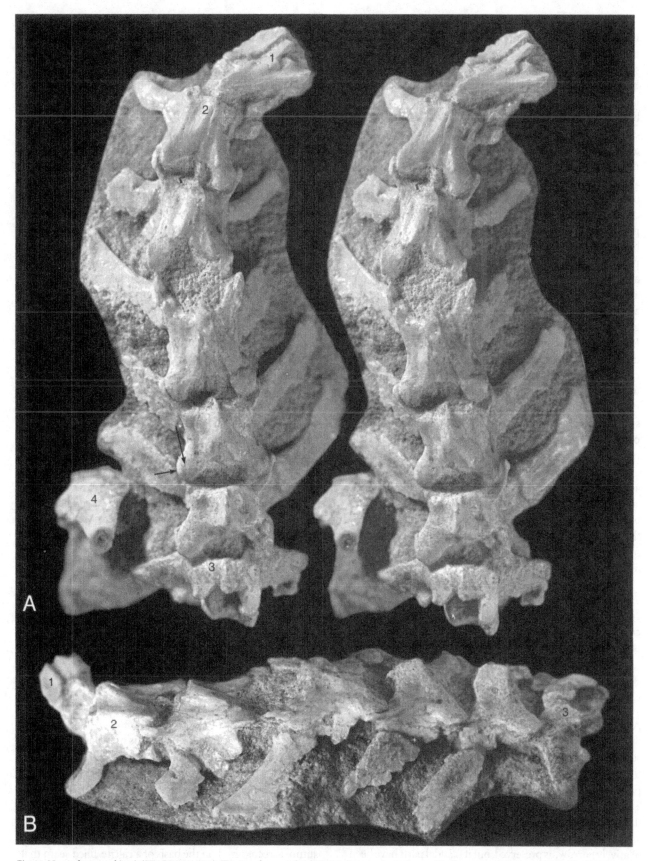

Fig. 11. Nemegtbaatar gobiensis (ZPAL MgM-I/81). □A. Lumbar vertebrae L2–L7, first sacral vertebra and broken part of left ilium of individual in Fig. 8A; dorsal view. □B. Same in left lateral view (ilium removed). 1 = L2; 2 = L3; 3 = S1; 4 = left ilium (broken); upper arrow in A = origin of m. sacrocaudalis dorsalis; lower arrow in A = origin of m. semispinalis dorsi. Both ×4; A stereo-pair.

The prezygapophysis is oval, large and faces dorsally and medially; the postzygapophysis is rounded and faces ventrally and laterally. Both pre- and postzygapophysis are slightly more massive on the posterior than on the anterior lumbar vertebrae. There is no metapophysis (proc. mamillaris). The anapophysis (proc. accessorius) is also lacking, but on the lateral sides of the pedicles on L4 and L5 (less certain on L3) there is a longitudinal lip in the place where an anapophysis occurs in other mammals, e.g., *Tachyglossus*, most marsupials, insectivores, carnivores and others. The spinous processes are more or less broken off on all the lumbar vertebrae near their bases, because of the damage done by preservation and preparation. It appears from the preserved parts (Fig. 11) and the preliminary drawings made before the preparation, that the spinous processes shift in position and change shape from L2 to L6 and were very long, as reconstructed in Fig. 36B$_2$. On the first two lumbars they are relatively small, arising from the cranial part of the arch and sloping cranially and dorsally. Caudally they become gradually more massive and more dorsally oriented. On L2–L5 there is a triangular surface that slopes caudally and downwards from the base of the spinous process and widens caudally. It is limited by lateral ridges that extend to the medial margins of the postzygapophyses and has a median ridge. This ridge extends dorsally along the midline of the spinous process. On consecutive vertebrae this surface becomes more ventrally oriented, aligned completely ventrally on L7.

The transverse processes (Figs. 8A, 11, 34A, 35, 36B, 39B, 40B) are relatively flat and very large; they extend craniolateroventrally and become broader towards L7. The complete left transverse process of L5 and the right one of L7 show that their extremities are extended longitudinally and project cranially as distinct processes, with a minute tuberlike thickening at the craniolateral corners.

It was possible to expose the ventral surface of L2 only. The length of the body is 5.1 mm, and there is a prominent ventral crest, partly broken. The body lateral to the crest is strongly concave. In cranial aspect the body of L2 is 3.9 mm wide and is truncated ventrally.

SACRUM
Figs. 8A, 11, 16G, H, 39A

An incomplete sacrum has been preserved in ZPAL MgM-I/81. S1 (Figs. 8A, 11) is in anatomical position with L7 and separated from S2; its body is shorter and wider than those of the lumbars. The spinous process is broken; the preserved part shows that it was apparently as large as on the lumbars. The transverse process is much shorter transversely and thicker than on the lumbar vertebrae; it is directed laterally and slightly ventrally. The transverse process appears to be relatively narrow, as preserved, but its caudal parts have been broken off on both sides. On the left side, the craniolateral corner of the transverse process is well preserved and bears

an inflation, similar to that in *Kryptobaatar*. This inflation fits in shape and size with the cranial part of the auricular surface preserved on the ilium. It is impossible to state whether two sacral vertebrae (as in *Kryptobaatar* and *Chulsanbaatar*) articulated with the ilium. It seems, however, that, as in *Kryptobaatar*, the ilium covered the transverse processes of the sacral vertebra (or vertebrae) dorsally. The deep incisure on the base of the transverse process, preserved on the left side, shows the position of the relatively large first dorsal sacral foramen. In ventral aspect S1 is relatively flat, inasmuch as the median ventral crest is lacking, developed only as a very faint ridge. Moreover the transverse processes are confluent with the body.

Only a small anterior fragment of the left side of S2 has been preserved (Figs. 8A, 16G, H), showing the relatively large (although broken) spinous process, the proximal part of the transverse process with the caudal margin of the first dorsal sacral foramen, and the fragment of the vertical wall of the left lamina. As the caudal part of S2 and the cranial part of S3 are missing, it is impossible to estimate the length of S2. In *Kryptobaatar* S2 is longer than S1, and this was probably the case in *Nemegtbaatar*. The transverse processes of both S3 and S4 are missing, only the bodies and the arches (without spinous processes) being preserved. Since S3 is broken, its length cannot be given; that of S4 is 5.4 mm. The pedicles of the fused arches of S3 and S4 arise dorsally, the laminae are arranged horizontally, and the intermediate sacral crest is rounded. The roots of the broken spinous processes arise from the caudal part of the vertebrae. The right postzygapophysis of S4 is preserved and is large and rounded, the articular surface being arranged horizontally. The ventral sides of the bodies of S3 and S4 are partly exposed. There is a prominent ventral crest, which widens at the caudal end of S4. At the longitudinal midpoint of S4, the crest widens and bears a longitudinal concavity. The sides of the bodies lateral to the crest are concave. The caudal epiphysis of the body of S4 is slightly concave, arranged slightly obliquely ventrocaudally.

Pectoral girdle and forelimb

Scapulocoracoid (Figs. 12, 13G–J, 28C). – A ventral fragment of a right scapulocoracoid is preserved in ZPAL MgM-I/81; the matrix has not been removed from the infraspinous fossa and from the base of the acromion because of the fragile state of the bone. The infraspinous fossa is only partly preserved and apparently widens dorsally. The glenoid fossa, the caudal border of which is partly missing, is pear-shaped. The coracoid and scapular parts form a broad angle of approximately 120°. The coracoid part is asymmetrical, L-shaped in cranial and caudal views, its ventral extremity being expanded medially to produce a prominent, rounded process. The coracoid suture is recognized on the basis of a calcite-filled seam that runs across the glenoid fossa. Along the cranial surface of the coracoid there is a rounded ridge with a minute tubercle.

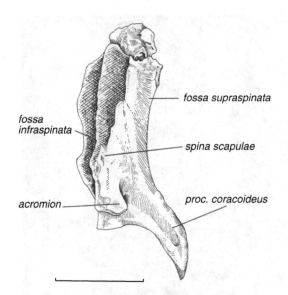

Fig. 12. □*Nemegtbaatar gobiensis* (ZPAL MgM-I/81). Camera lucida drawing of right scapulocoracoid of individual in Fig. 8A (see also Fig. 13G–J), lateral view. Scapular spine broken, and ventral part of acromion (seen in Fig. 13G–J) is not shown. Scale bar 5 mm.

This ridge continues for a short distance on the scapula as the cranial border, although it is less prominent. Beyond the cranial scapular border the ridge continues dorsally as a very sharp keel, strongly reflected laterally on the dorsal end.

The shoulder blade (Fig. 13H–J) is narrow and relatively flat ventrally. A weak thickening extends on the blade along the cranial border and another one longitudinally along the highest convexity of the blade, on the subscapular fossa. Dorsally the blade widens and becomes convex. The subscapular fossa is convex, except for the part lying along the keel of the cranial border, where there is a small concavity (number 11 in Fig. 13I, J). The caudal margin just above the glenoid fossa is gently incurved. On the ventral part of the medial side of the blade, lying in prolongation of the coracoid edge, there is a ridge that extends in an arch towards the ventral margin (number 13 in Fig. 13H). The ridge is more prominent ventrally than dorsally, with a shallow pit behind it, just above the glenoid fossa.

The craniolateral part of the blade is incomplete (Figs. 12, 13I), but the matrix preserved in the caudal dorsal part apparently corresponds to the shape of the missing bone. The preserved part of the spine is short, limited mostly to the ventral part of the scapula, and has a wide base. Dorsally it was confluent with the missing caudal margin of the craniolateral blade, thus the spine as a whole is arranged obliquely across the blade. The dorsal part of the craniolateral blade between the keel of the cranial border and the dorsal prolongation of the spine is concave and is recognized here as an incipient supraspinous fossa (Figs. 12 and number 9 in Fig. 13I). The spine extends ventrally, turning slightly caudally and continues as a peg-like acromion process; this is oriented

ventrally, slightly cranially, parallel to the ridge on the outer surface of the glenoid fossa. At the cranioventral end of the spine, on the ventral edge of the scapulocoracoid there is a depression (number 12 in Fig. 13I).

Humerus (Figs. 13A–E, 31B). – The proximal part of the right humerus preserved in ZPAL MgM-I/81 (Fig. 13A–E) is robust. As is characteristic of multituberculate humeri, the greater tubercle (tuberculum majus) is higher and lies closer to the head than does the lesser tubercle (tuberculum minus). The intertubercular (bicipital) groove (sulcus intertubercularis) is deep, especially on the side of the greater tubercle. It narrows slightly distally and then widens again. The posterior crest, running in proximal prolongation of the ectepicondylar flange to the head (Kielan-Jaworowska & Dashzeveg 1978) is not very prominent. As is also characteristic of multituberculate humeri, the head overhangs the shaft dorsally and the crest of the lesser tubercle (teres tuberosity, tuberositas teres major) is crescent-shaped, not very prominent (number 5 in Fig. 13B, D). The damaged distal fragment of the same humerus does not show characters that would differentiate it from other multituberculate humeri (see distal end of the unidentified multituberculate humerus from the Hell Creek Formation, Montana, figured for comparison in Fig. 14B). The entepicondylar region of the *Nemegtbaatar* humerus, showing uncertain muscle scars, is figured in Fig. 30B. Of all the proximal multituberculate humeri known (see Krause & Jenkins 1983 and Kielan-Jaworowska & Dashzeveg 1978 for reviews), the humerus of *Nemegtbaatar* resembles that of a taeniolabidoid, fam., gen. et sp. indet. (ZPAL MgM-I/165) from the Djadokhta Formation (Kielan-Jaworowska 1989). The *Nemegtbaatar* humerus is only slightly smaller than ZPAL MgM-I/165 and differs from it in that the humeral head is somewhat higher and more rounded, the intertubercular groove is not as deep and the deltopectoral crest less prominent. It is larger and more robust than in *Tugrigbaatar* (Kielan-Jaworowska & Dashzeveg 1978), but much smaller and less robust than that of ?*Lambdopsalis* (Kielan-Jaworowska & Qi 1990). Also the teres tuberosity (not preserved in ZPAL MgM-I/165), in *Nemegtbaatar* is less prominent than in ?*Lambdopsalis.*

Radius and ulna (Figs. 13F, 14A). – A partial left ulna with a broken olecranon and missing the distal extremity has been preserved together with the proximal part of the radius in ZPAL MgM-I/81 (Fig. 14A). The ulna is strongly compressed laterally; its mediolateral diameter at the level of the radial head is 1.0 mm, and the dorsoventral diameter is 1.9 mm. The ulnar shaft slightly decreases in craniocaudal diameter distally and becomes roughly triangular in cross section. Along the medial side of the shaft there extends a longitudinal fossa, bordered by a prominent crest for the interosseus ligament. See also the proximal part of an unidentified multituberculate ulna from the Hell Creek Formation, Montana, figured for comparison (Figs. 14C–E, 32B, 33B).

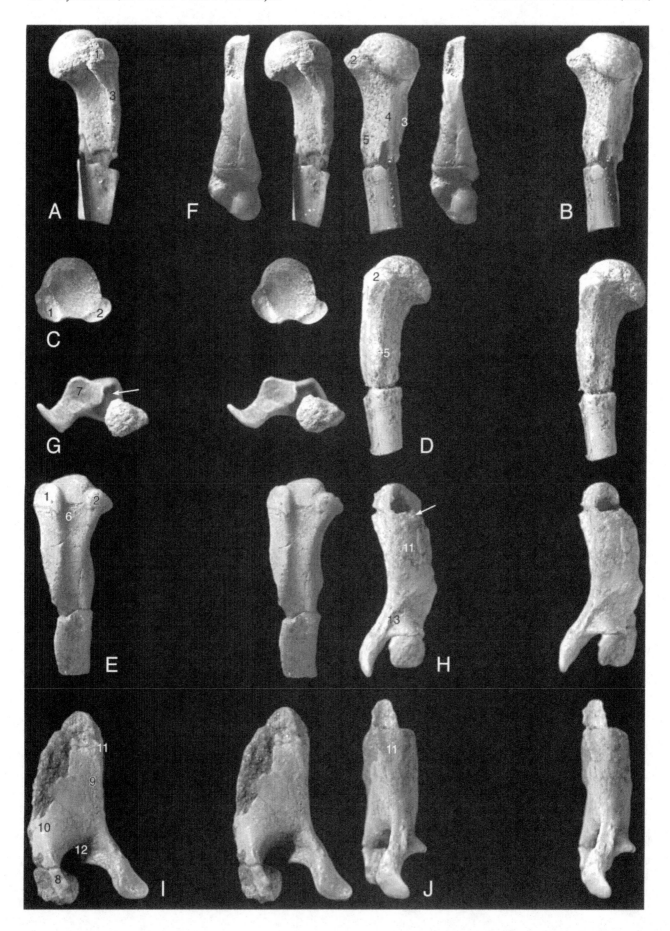

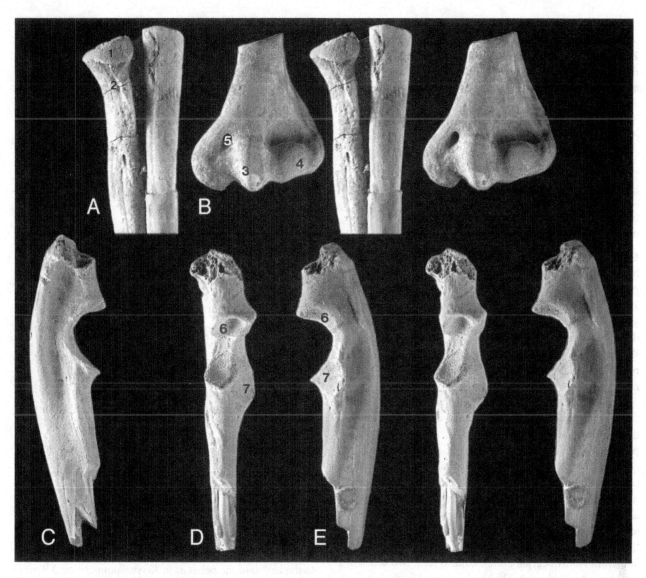

Fig. 14. □A. *Nemegtbaatar gobiensis* (ZPAL MgM-I/81). Proximal part of left radius and ulna (olecranon broken) of individual in Fig. 8A, lateral view. Cranial margin of radius shifted slightly medially; caudal margin of ulna also shifted medially. □B. Distal part of left humerus of unidentified multituberculate (AMNH 118267), ventral view, Bug Creek Anthills site, Hell Creek Formation, Montana. □C, D, E. Left ulna of unidentified multituberculate (AMNH 118505), horizon and locality as in B. Proximal part in medial, cranial and lateral views. 1 = articular circumference; 2 = tuberosity for insertion of m. biceps brachii; 3 = ulnar condyle; 4 = radial condyle; 5 = entepicondylar foramen; 6 = articular surface for ulnar condyle; 7 = radial notch. A ×6; B–E ×4; all except C stereo-pairs; all coated with ammonium chloride.

Fig. 13 (opposite page). *Nemegtbaatar gobiensis* (ZPAL MgM-I/81). Isolated parts of skeleton in Fig. 8A. □A–E. Proximal part of right humerus. □A. Lateral view. □B. Dorsal view. □C. Proximal view. □D. Medial view. □E. Ventral view. □F. Distal part of right radius, associated with incomplete scaphoideum (to the right) and lunatum, in cranial view. □G–J. Right scapulocoracoid. □G. Ventral view. □H. Craniomedial view. □I. Craniolateral view. □J. Cranial view. 1 = greater tubercle; 2 = lesser tubercle; 3 = deltopectoral crest; 4 = posterior crest; 5 = teres tuberosity; 6 = intertubercular groove. 7 = glenoid fossa; 8 = acromion; 9 = incipient supraspinous fossa; 10 = ventral part of the spine; 11 = subscapular fossa; 12 = origin of m. infraspinatus; 13 = ridge for fourth (most cranioventral) aponeurosis of m. subscapularis; arrows in G and H point to infraspinous fossa, filled with matrix. All ×4; stereo-pairs.

On the proximal part of the radius (Fig. 14A), there is a roughly semilunar, obliquely directed medial facet (articular circumference, circumferentia articularis), for articulation with the ulna. Distal to that facet there is a triangular tuberosity for further contact with the ulna, medial to which there extends a groove for muscular insertion (see 'Myological reconstructions'). The shaft is craniocaudally compressed. The radial head is mediolaterally elongated; its mediolateral diameter is 1.7 mm, and the diameter of the ulna at the level of the radial head is 2.3 mm.

The distal end of the right radius has been preserved together with two carpal bones, exposed in cranial view in ZPAL MgM-I/81 (Fig. 13F). The middle part of the shaft of the radius is 1.3 mm along the craniocaudal axis and 1.9 mm along the transverse axis. The shaft widens strongly distally, and the distal epiphysis is 3.3 mm wide. Along the cranial side of the shaft there extends a longitudinal groove that narrows distally. On the distal end of this groove is a longitudinal crack. The distal epiphysis is distinguished from the shaft owing to the deeply excavated suture between the two areas. It bears a prominent, triangular styloid process. The caudal side of the radius is concave. In the same piece of rock fragments of broken metacarpals (not figured) have been preserved.

Carpus. – ?Right scaphoideum (radial carpal bone) and lunatum (semilunare, intermedial carpal bone) have been preserved in association with the above described radius of ZPAL MgM-I/81 (Fig. 13F), exposed in cranial view. The scaphoideum is damaged, roughly oval, elongated proximo-distally and situated distal to the styloid process (possibly slightly displaced laterally). The lunatum is much smaller, roughly circular and situated distal to the lateral part of the radius.

Two isolated carpal bones (Fig. 15) have been preserved within one piece of rock, some distance from the radius described above. They are tentatively identified as right carpal bones and have been exposed in oblique distal view. The one on the right side of Fig. 15 is possibly the triquetrum (ulnar carpal bone). It is transversely elongated, about 1.75 mm wide, and has a slightly convex proximal facet for articulation with the ulna and a small medioproximal facet for articulation with the lunatum. On the mediodistal side there is a strongly concave facet for articulation with the hamatum (fourth carpal bone). On this facet two nutrient foramina are clearly seen. The bone preserved on the left side in the same piece of rock apparently belongs to the distal row of the carpals. The shape indicates that it might be a displaced pisiform (os pisiforme); it is too large to be the hamatum, as it does not fit the facet on the triquetrum for articulation with the hamatum. It might also be the trapezoideum (second carpal bone) or praepollex. In Fig. 15 it is arranged obliquely. It is strongly elongate and measures about 1.9 mm across its longer diameter. It has a convex proximal facet (facing upward and to the left in Fig. 15) and a concave distal facet.

Pelvis and hind limb

Pelvis (Figs. 8A, 16D–F, 37A, 39, 41). – The caudal parts of both halves of the pelvis are preserved in ZPAL MgM-I/81, displaced ventrally relative to the lumbar vertebrae; the anterior part of the right ilium (Fig. 16D, E) is preserved separately in front of the lumbar vertebrae. The estimated length of the pelvis is about 34 mm. The wing of the ilium is similar to that in *Kryptobaatar* but differs in having a more rounded

Fig. 15. Nemegtbaatar gobiensis (ZPAL MgM-I/81). Isolated right carpal bones, in oblique distal view, of skeleton in Fig. 8A. 1 = ?triquetrum; 2 = displaced ?pisiform, ?trapezoideum or ?praepollex. Stereo-pair; ×8.

caudal ventral iliac spine. The auricular surface consists of a large, rounded caudal part that projects ventrally, forming the medial wall of the caudal ventral iliac spine, and a smaller cranial part separated from the caudal one by a weak vertical ridge. The caudal part is delimited posteriorly by a distinct vertical ridge. The line delimiting the gluteal surface dorsally forms a sharp ridge (much more prominent than in *Kryptobaatar*) that extends caudally to a notch in front of the acetabulum. The boundary between the body of the ilium and the pubis cannot be recognized with any certainty. It possibly extends above the sharp ridge that runs cranially from the ventral margin of the acetabulum (Fig. 16F). The cranial border of the acetabulum is thickened and differs from that in *Kryptobaatar* in having a more prominent, nodule-like thickening, with an oval pit for m. rectus femoris in front of it (arrow in Fig. 16F). The acetabulum is 3.4 mm long, 3.2 mm high (both inner dimensions) and 4.6 mm long (including margins).

The obturator foramen has the same proportions relative to the length of the pelvis as in *Kryptobaatar*, being 3.9 mm long. Its ventral margin (preserved on the right side) has a rather unusual shape for mammals, with a small rounded process that protrudes into the foramen (Fig. 16F). Although the caudal part of the ventral margin of the foramen to the rear of the process is slightly distorted, the well-preserved process does not seem to be an artefact. The caudal border of

Fig. 16. Nemegtbaatar gobiensis (ZPAL MgM-I/81), isolated parts of skeleton in Fig. 8A. □A, B, C. Proximal part of right femur in ventral, dorsal and medial views. □D, E. Anterior part of right ilium in dorsal and medial views. □F. Posterior part of right pelvis, lateral view. □G, H. Incomplete sacral vertebrae S2–S4 in dorsal and left lateral views. (For S1 see Fig. 11). 1 = trochanteric fossa; 2 = posttrochanteric fossa; 3 = lesser trochanter; 4 = rugose area on greater trochanter for insertion of gluteal musculature; 5 = subtrochanteric tubercle; 6 = postobturator notch (or foramen); 7 = indentation on margin of pubis, possibly for epipubic bone; 8 = acetabulum; 9 = obturator foramen; 10 = S2; 11 = S4; left arrows in D and right arrows in E = origin of dorsal part of m. gluteus medius; left upper arrows in E = origin of m. longissimus dorsi; left lower arrows in E = auricular surface (two parts), forming medial wall of caudal ventral iliac spine; arrow in F = pit for m. rectus femoris. All ×4; all except D and E stereo-pairs.

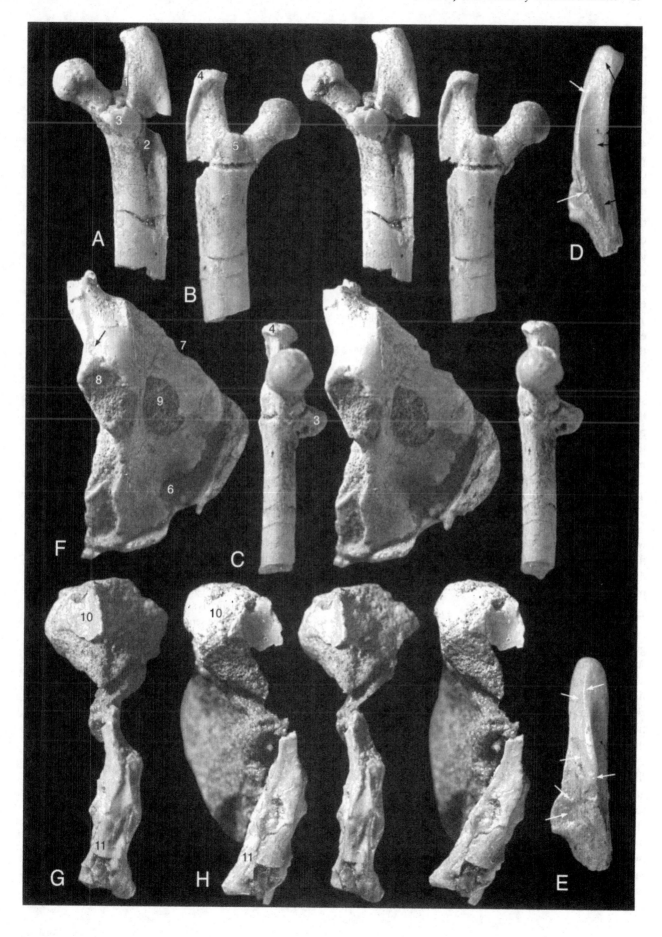

the acetabulum is raised in a similar way as the cranial, with a prominent nodule-like structure and a ventrocaudal pit behind it. The dorsal margin of the ischium is thickened (broken on the right side and showing only the medial wall caudally) and directed dorsally forming a concave surface. The ischial tuber is not preserved. The ventral margins of the ischia, fused in other genera to form a keel, are broken off, and left and right halves of the pelvis are slightly displaced with regard to each other; the keel has not been preserved. Caudal to the obturator foramen there is a postobturator notch (possibly a foramen), preserved as a deep indentation on the ventral margin. This is relatively larger and higher than in *Kryptobaatar,* and we cannot determine whether it opened into the pelvic cavity. The lateral surface of the ischium is concave, especially in the area dorsal and cranio-dorsal to the postobturator notch. From the cranioventral corner of the postobturator notch, two faint blunt ridges extend craniodorsally and reach the obturator foramen. The area of the apparent fusion of the ischium and pubis is elevated, extending from the ventral margin of the pelvis to the above described process at the ventral margin of the obturator foramen.

The pubis is relatively small, with an undulating lateral surface. There is a tubercle on the ventral margin in front of the presumed line of fusion with the ischium. In front of the tubercle, at a point midway along the length of the pubis, there is a deep, rounded indentation 1.8 mm long and 0.4 mm high, possibly for articulation with the epipubic bone. This is very different from the shallow, elongated indentation in *Kryptobaatar.* Dorsal to the indentation there is another one, very small and shallow, which may be due to distortion.

The pelvis of *Nemegtbaatar* appears to be relatively more widely open than in *Kryptobaatar;* the pubes and ischia meet their counterparts possibly at wider angles. Because of the distortion caused by crushing, these angles cannot be estimated.

Femur (Figs. 8A, 16A–C, 17B–E, 41C, 42). – The femur in *Nemegtbaatar* differs from that in *Kryptobaatar* in having a greater trochanter that is relatively more robust and more strongly protruding proximally over the head (Fig. 16A–C). There is a ridge that delimits the triangular rugose area for insertion of the gluteal musculature (on the caudal part of the ridge). On the cranial part of this ridge originated the apo-neurosis for m. vastus lateralis. This ridge is better developed in *Nemegtbaatar* than in *Kryptobaatar* in that it is higher, exhibits fine crenulations and has a pitted furrow along its distal and medial sides. In lateral view the rugose area for attachment of gluteals is longitudinally divided by a faint furrow, at the ventral side of which inserts m. gluteus medius (Fig. 42C). In proximal view this area is divided by a furrow directed from the ventromedial to the dorsolateral side into two semicircular parts, the lateral of which is larger. The medial part is subdivided by a shallow, dorsally concave furrow into a larger, oval, ventral part and a crescent-shaped

dorsal one. The lesser trochanter is very prominent, similarly built as in *Kryptobaatar.*

The femoral head is more sharply delimited from the neck than in *Kryptobaatar.* The heart-shaped rugosity on the head, representing the fovea capitis femoris, which was noted by Jenkins & Krause 1983 as characteristic for multi-tuberculates studied by them, is hardly recognizable. A faint ridge extends along the distal part of the neck, between the head and the base of the greater trochanter. The trochant-eric fossa is relatively shallow; it extends as a narrow groove onto the distal part of the greater trochanter and narrows proximally (number 1 in Fig. 16A). The post-trochanteric fossa, which is shallow in *Kryptobaatar,* is very deep here, and the gluteal crest is much more prominent, forming a keel that is 2.3 mm wide proximally. In dorsal aspect, at the divergence of the neck and the greater trochanter, there is a prominent subtrochanteric tubercle, similarly developed as in *Kryptobaatar.*

At the distal extremity (Fig. 17B–E), the lateral condyle (preserved in ZPAL MgM-I/110) protrudes less over the shaft than in *Kryptobaatar,* and the trochlea is shifted later-ally, closer to the lateral condyle. In cranial view the medial ridge of the trochlea is more prominent than the lateral one.

?Parafibula. – A bone fragment that may represent a neck of the parafibula, has been preserved in the knee joint in ZPAL MgM-I/110 (Fig. 17F, H, I).

Tibia (Figs. 17F–H, 43). – In ZPAL MgM-I/110 the left tibia is 19.4 mm long, and its proximal extremity is 5.5 mm wide in cranial view. This tibia is possibly slightly deformed, being compressed distally, and as a consequence more strongly bent laterally than it originally was. The proximal end is asymmetrical, more triangular than oval, with a rounded, caudally situated spine. In this respect it differs from *Krypto-baatar* and *?Stygimys* (Krause & Jenkins 1983, Fig. 23E), in which taxa the proximal end of the tibia is roughly oval and the caudal prominence is small. In *Nemegtbaatar* the caudal border encircles a large semicircular prominence that abuts against the fibula. Laterally the proximal articular surface bears a prominent hook-like process (Fig. 17F, H, I). The proximal articular surfaces are more prominent than in *Kryptobaatar,* inasmuch as the concavities of the medial and lateral articular facets are deeper in *Nemegtbaatar.* The tibial

Fig. 17. □A. *Chulsanbaatar vulgaris* (ZPAL MgM-I/99b), Red beds of Khermeen Tsav, Khermeen Tsav II, Gobi Desert, Mongolia. Partial left knee joint; femur to the right. Lateral view. □B–I. *Nemegtbaatar gobiensis* (ZPAL MgM-I/110), horizon and locality as in A. □B–E. Distal part of left femur. □B. Dorsal (slightly lateral) view. □C. Lateral view. □D. Medial view. □E. Ventral view. □F–I. Left tibia and fibula in lateral, medial, caudal and cranial views. 1 = tibia; 2 = fibula, 3 = hook-like lateral process of fibula; 4 = medial condyle; 5 = lateral condyle; 6 = hook-like lateral process of tibia; 7 = fragment of a ?parafibula; 8 = tibial tuberosity; 9 = cranial crest; arrow in G = pit for lig. collaterale mediale; arrow in I = fusiform area for insertion of m. semimembranosus anterior. All ×4; stereo-pairs.

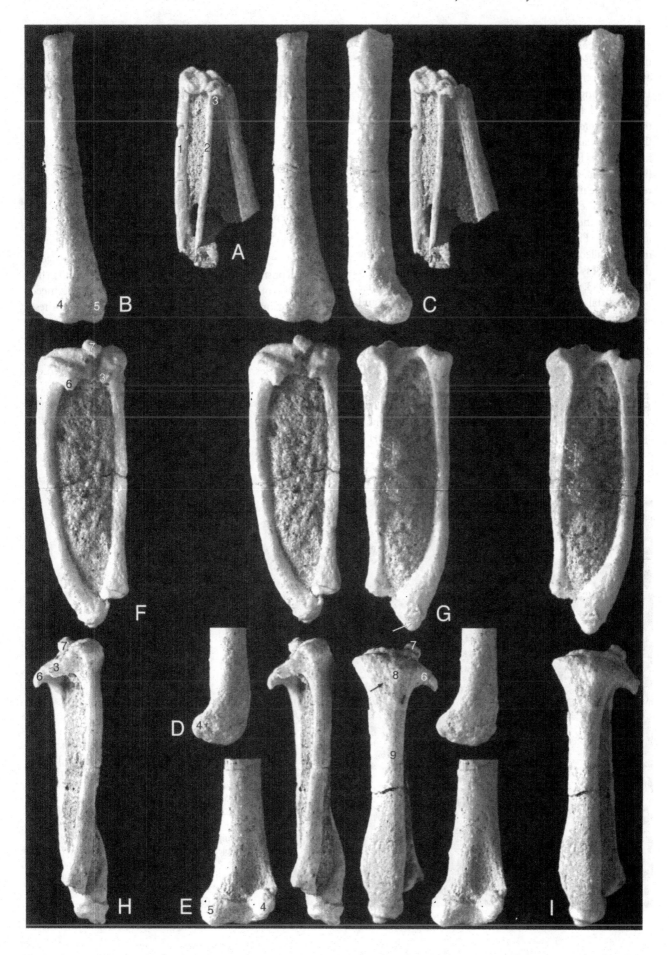

tuberosity is distinct but flattened. The cranial (anterior) crest (number 9 in Fig. 17I) is sharp proximally. At the middle of the length, the shaft is 2.5 mm wide transversely and 1.3 mm craniocaudally. The tibia in ZPAL MgM-I/110 is 19.4 mm long (compared with 18.8 mm in *Kryptobaatar*, ZPAL MgM-I/41), but otherwise it is much more robust than in *Kryptobaatar*. Extending along the proximal part of the medial wall, cranial and parallel to the cranial crest, is a fusiform area for insertion of m. semimembranosus anterior (arrow in Fig. 17I and Fig. 43C, D). These structures are not recognizable in *Kryptobaatar*.

The distal end comprises the epiphysis, distinctly separated from the shaft, which indicates the young age of the individual (Fig. 17F–I). The distal lateral condyle and medial malleolus are poorly recognizable, but apparently they were less prominent than in ?*Eucosmodon* (Krause & Jenkins 1983, Fig. 22A, B). In ?*Eucosmodon* we recognize on the distal surface of the tibia two condyles (lateral and small medial) and the medial malleolus (Fig. 56E, F), but in *Nemegtbaatar* the medial condyle apparently was not developed, as is characteristic also of an unidentified tibia from the Hell Creek Formation (Fig. 55A; see also 'Anatomical comparisons').

Fibula (Figs. 17F–I, 43A). – The length of the fibula in ZPAL MgM-I/110 cannot be given because the distal epiphysis (including the distal caudal tuberosity; see *Kryptobaatar*) is not preserved. The head is 2.2 mm long craniocaudally and 4.1 mm wide (including the hook-like process) in caudal view (Fig. 17H). The facet for articulation with the parafibula is very large. Along the caudal wall of the shaft there extends a sharp crest, not exposed in *Kryptobaatar*.

Genus *Chulsanbaatar* Kielan-Jaworowska, 1974

Chulsanbaatar vulgaris Kielan-Jaworowska, 1974

(Figs. 17A, 18–25, 37B)

Material. – We tentatively assign to *Chulsanbaatar vulgaris* all small postcranial multituberculate fragments found in the Barun Goyot Formation and in the Red beds of Khermeen Tsav. It cannot be excluded, however, that some of these skeletons, especially those not associated with skulls, belong to another, as yet unrecognized, multituberculate taxon.

Barun Goyot Formation: ZPAL MgM-I/61, Khulsan: skull with dentaries, damaged C2–T2, cervical ribs preserved on C2 and on C5, an isolated cervical rib possibly of C7, a damaged fragment of the scapulocoracoid in articulation with damaged proximal fragment of the right humerus. ZPAL MgM-I/111, Nemegt, Eastern Sayr: skull with den-

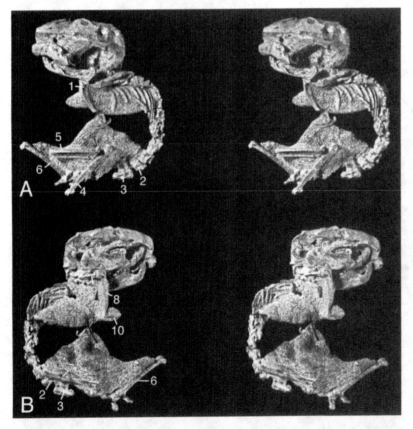

Fig. 18. □A, B. *Chulsanbaatar vulgaris* (ZPAL MgM-I/83), Red beds of Khermeen Tsav, Khermeen Tsav II, Gobi Desert, Mongolia. Flattened skull, dentaries and fairly complete, damaged postcranial skeleton without caudal vertebrae and distal parts of the limbs, as preserved, after some preparation. □A. Left lateral view. □B. Right lateral view. 1 = left humerus; 2 = L7; 3 = incomplete sacrum; 4 = left femur; 5 = left tibia (incomplete fibula parallel to it); 6 = right femur; 7 = right ilium; 8 = right humerus; 9 = right scapulocoracoid; 10 = proximal part of right radius and ulna. See also Figs. 22 and 23. Some elements figured in Fig. 22 are not exposed. Both ×1; stereo-pairs.

taries, atlas, C2–C4, the body of C5, ventral parts of both scapulocoracoids in articulation with proximal parts of humeri, distal part of the right radius in articulation with the scaphoideum. ZPAL MgM-I/138, Khulsan: skull with dentaries associated with broken fragments of the long bones. ZPAL MgM-I/140, Khulsan: skull with dentaries, associated with broken fragments of ribs and long bones. ZPAL MgM-I/143, Khulsan: anterior part of the skull, dentaries, poorly preserved fragments of cervicals and long bones; ZPAL MgM-I/149, Khulsan: left dentary associated with incomplete left pelvis. ZPAL MgM-I/151, Khulsan: isolated upper teeth and fragments of edentulous dentaries, radius, fragments of metatarsals, incomplete ilia and ischia with partial femora in acetabula; this specimen may not belong to *Ch. vulgaris*. ZPAL MgM-I/170a, Nemegt: postcranial fragments only: L6, L7, partial sacrum, parts of both ilia, proximal part of the right femur.

Red beds of Khermeen Tsav, Khermeen Tsav II: ZPAL MgM-I/83: skull, dentaries, bodies of C2–T2, bodies of ?T7–T13, L1–L7, incomplete sacrum, numerous broken ribs, fragmentary right and left scapulocoracoids, right and left humeri, each in two parts, proximal parts of right and left radii and ulnae, right ilium, left ischium, right and left femora, left tibia and fibula (without the proximal part). ZPAL MgM-I/84: skull cut on the microtome (Kielan-Jaworowska *et al.* 1986; Hurum 1992, 1994), fragments of the cervical vertebrae and humerus, fragments of long bones; all the postcranial elements are badly damaged. ZPAL MgM-I/85 only postcranial skeleton: damaged sacrum, partial ilia and ischia, proximal part of the left femur, the femoral head and greater trochanter of the right femur; both femoral heads are in acetabula. ZPAL MgM-I/99a and ZPAL MgM-I/99b found together in the same piece of rock: 99a is a very young individual; the skeleton consists of two dentaries, fused,

incomplete pubes and ischia in articulation with a partial left and a complete right femur, the latter in articulation with the right tibia and fibula, patella and the parafibula; this specimen may not belong to *Ch. vulgaris*. Specimen 99b consists of an incomplete right femur in articulation with the tibia and fibula, the shaft of the left femur and incomplete left pes that includes the tarsus and five metatarsals. ZPAL MgM-I/108: skull, dentaries, five cervicals, exposed in ventral view, partial left scapulocoracoid in articulation with the proximal part of the humerus, partial left femur, broken ribs. ZPAL MgM-I/145: skull with a fragment of the left dorsal arch of the atlas, both dentaries, the proximal part of a damaged left humerus, a partial hyoid arch and some unidentified bones below it.

Skull
Figs. 18, 19B and 20C

Numerous skulls of *Chulsanbaatar* were described by Kielan-Jaworowska (1974), Kielan-Jaworowska *et al.* (1986), and Hurum (1992, 1994). In the present paper we figure only three skulls found in association with postcranial fragments.

Hyoid apparatus
Fig. 19

In ZPAL MgM-I/145 an incomplete hyoid apparatus has been preserved below and somewhat caudal to the skull (Fig. 19). The preserved part consists of two stylohyoid bones that form an angle of 80°, almost touching each other. The basihyoid and the rest of the hyoid apparatus have not been preserved. Both stylohyoid bones have been pushed caudally and turned a little more than 90° ventrocaudally, so that in Fig. 19B their point of juncture is directed caudally. When

Fig. 19. Chulsanbaatar vulgaris (ZPAL MgM-I/145), Red beds of Khermeen Tsav, Khermeen Tsav II, Gobi Desert, Mongolia. □A. Partial hyoid apparatus separated from skull in B, dorsal view. □B. Almost complete skull with both dentaries, showing partial hyoid apparatus, rotated ventrocaudally from its natural position. 1 = left stylohyoid bone; 2 = right stylohyoid bone. Stylohyoid bones appear smaller in B than in A because of oblique position of hyoid apparatus in B. Both ×4; A stereo-pair.

the piece of rock containing the stylohyoid bones was separated from the skull, the photograph in Fig. 19A was taken, showing the hyoids in dorsal aspect. Both hyoids are cracked. The right one, which is more complete and better preserved, is 5.8 mm long (the length of the skull is 18.6 mm). The stylohyoids are convex dorsally, slightly sigmoidal and slightly expanded at both ends.

Axial skeleton
Figs. 18, 20, 21

CERVICAL VERTEBRAE AND CERVICAL RIBS

Atlas. – In ZPAL MgM-I/111 (Fig. 20C) the atlas is complete, preserved between the skull and the axis, its ventral arch being obscured by the axis and the following cervicals. As the bones are strongly cemented and cracked, the atlas was not removed from its original position. In caudal view the atlas is 6.6 mm wide (including the transverse processes), and its height in lateral view is 3.3 mm. The right and left laminae are separated at the top. The transverse process, well preserved on the left side, is relatively small, roughly rectangular, transversely elongated and with rounded tips. The transverse foramen is absent. The left fragment of the atlas preserved in ZPAL MgM-I/145 has a transverse process and a caudal articular surface that is almost flat.

Axis. – The arch is incomplete in all three specimens (see 'Material'), and the size of the spinous process is not known. In ZPAL MgM-I/61 (Fig. 20A), on the displaced axis the left cervical rib is preserved in articulation with the axis. It is

relatively large and flat, rounded caudally and directed obliquely caudally, more ventrally than horizontally. Although the suture with the transverse process is hardly discernible, we identify this structure tentatively as a cervical rib, as it is similar in position and shape to that of *Ornithorhynchus* (Lessertisseur & Saban 1967a).

C3–C7. – These vertebrae are preserved in several specimens (see 'Material') and are badly damaged. The laminae, partly preserved in ZPAL MgM-I/61 (Fig. 20B) and ZPAL MgM-I/111 are directed dorsomedially, but possibly less dorsally than in *Nemegtbaatar*. Otherwise they do not reveal features that would differentiate them from those of *Nemegtbaatar*. A badly damaged apparent right cervical rib, associated with C5 (Fig. 20B), is preserved. In lateral view this element is flat, roughly oval, narrowing caudally, directed horizontally and slightly ventrally. Posterior to the fifth cervical rib, another apparent cervical rib has been preserved, possibly the seventh, obscured in Fig. 20B. This one differs from the fifth one in having a less curved outline.

THORACIC VERTEBRAE AND RIBS

The anterior thoracic vertebrae preserved in two specimens (ZPAL MgM-I/61 and ZPAL MgM-I/83) are badly damaged and are therefore not described. The seven posterior thoracics preserved cranial to L1–L7 in ZPAL MgM-I/83 (Fig. 18) are also damaged. Only the bodies are preserved, but they are obscured by broken ribs in several places. The number of thoracic vertebrae is not known in multituberculates. We tentatively assume that there were 12 thoracics in *Chulsanbaatar*, and we designate the seven posterior thoracics as T6–

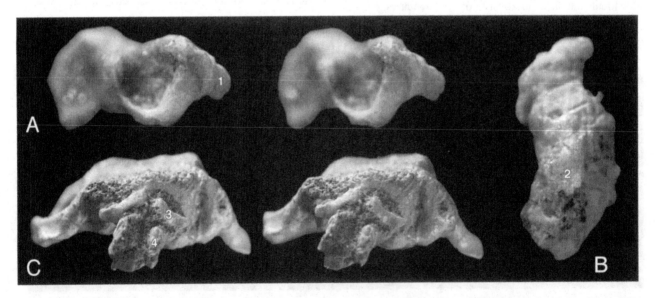

Fig. 20. Chulsanbaatar vulgaris. □A, B. ZPAL MgM-I/61, Barun Goyot Formation, Khulsan, Gobi Desert, Mongolia. □A. Damaged axis in cranial view. □B. C2–T1 of same specimen (C2 is displaced and poorly seen) in left lateral view. Apparent seventh cervical rib has been displaced and is not seen in this view. □C. ZPAL MgM-I/111, horizon and locality as in A. Skull and anterior cervical vertebrae in caudal view. 1 = apparent left cervical rib of the axis; 2 = fifth cervical rib; 3 = atlas, showing transverse processes; 4 = damaged C2–C3. A, B ×8; C ×4; A, C stereo-pairs.

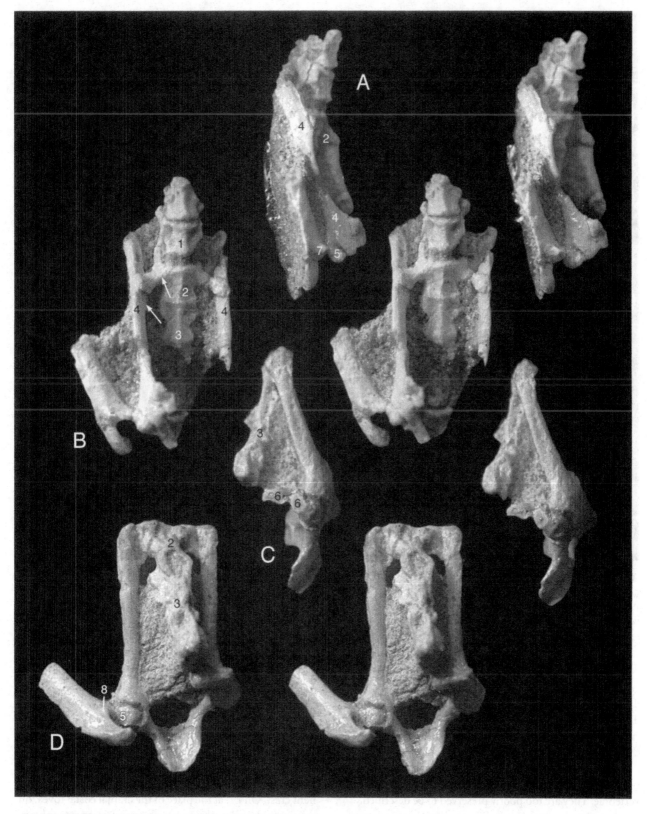

Fig. 21. Chulsanbaatar vulgaris. □A, B. ZPAL MgM-I/170a, Barun Goyot Formation, Nemegt, Gobi Desert, Mongolia. L6, L7, partial sacrum, partial ilium, and proximal part of right femur. □A. Left lateral view. □B. Ventral view. □C, D. ZPAL MgM-I/85. Red beds of Khermeen Tsav, Khermeen Tsav II, Gobi Desert, Mongolia. Damaged sacrum, ilia, and ischia, proximal part of left femur, femoral head and greater trochanter of right femur; both femoral heads are in acetabula. □C. Right lateral view (dorsal is to the left). □D. Dorsal view. 1 = L7; 2 = S1; 3 = S2; 4 = ilium; 5 = femoral head in acetabulum; 6 = broken right femur; 7 = lesser trochanter; 8 = subtrochanteric tubercle; upper arrow in B = crest on transverse process of S1; lower arrow in B = facet on right ilium for articulation with the transverse process of S2. All ×4; stereo-pairs.

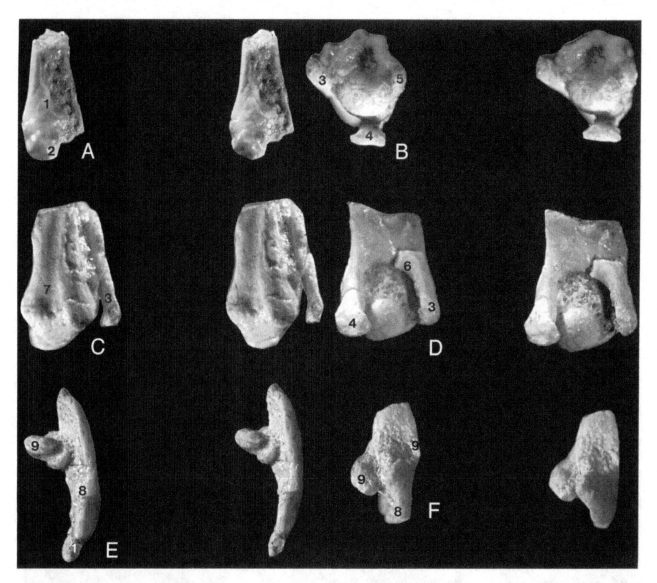

Fig. 22. Chulsanbaatar vulgaris. □A–D. ZPAL MgM-I/111, postcranial elements associated with skull in Fig. 20C. □A. Distal end of right radius, in articulation with damaged scaphoideum, ventral view. □B–D. Same individual, proximal end of left humerus in articulation with ventral end of scapulocoracoid. □B. Proximal view of humerus, ventral view of scapulocoracoid. □C. Humerus and acromion in ventral view. □D. Scapulocoracoid in cranial view, humerus in dorsal view. □E, F. ZPAL MgM-I/83. □E. Lateral view of ulna and radius of individual in Fig. 18, in articulation with distal end of poorly preserved humerus. □F. Same specimen, proximal view of ulna and dorsal view of humerus. 1 = radius; 2 = scaphoideum; 3 = acromion; 4 = coracoid process; 5 = greater tubercle; 6 = origin of m. infraspinatus; 7 = intertubercular groove; 8 = ulna; 9 = distal end of humerus. All ×8; stereo-pairs.

T12. The lengths of their bodies are: T6 1.7 mm, T7 1.8 mm, T8 1.8 mm, T9 2.0 mm, T10 2.1 mm, T11 2.1 mm and T12 2.3 mm. Poorly preserved costal foveae are preserved on some of the bodies, but the caudal costal foveae are uncertain on T12.

Fragments of ten ribs have been preserved on the left side of ZPAL MgM-I/83 (Fig. 18), in association with the thoracic vertebrae. They are too poorly preserved to be described.

LUMBAR VERTEBRAE

The lengths of the bodies in ZPAL MgM-I/83 (Fig. 18) are: L1 2.4 mm, L5 2.4 mm, L6 2.7 mm and L7 1.9 mm (other bodies are distorted). In ZPAL MgM-I/170a, L6 is 2.7 mm and L7

1.9 mm long. The spinous processes on L1–L3 (in ZPAL MgM-I/83) are directed mostly cranially, only slightly dorsally, and are possibly rounded at the extremities and overlap each other. Those on the successive vertebrae rise more steeply dorsally. The pre- and postzygapophyses are relatively large. In ventral aspect all the lumbars have a prominent crest. On L1, at the place of the cranial costal fovea, on the right side, below the poorly preserved postzygapophysis, a small rounded piece of bone is preserved that might be the broken-off head of the last rib. If so, the vertebra recognized by us as L1 is indeed the first lumbar, which indicates that there are seven lumbars in *Chulsanbaatar*.

Fig. 23. Chulsanbaatar vulgaris (ZPAL MgM-I/83). A–D. Proximal end of left humerus of individual in Fig. 18. □A. Lateral view. □B. Dorsal view. □C. Proximal view. □D. Medial view. □E. Ventral view. 1 = greater tubercle; 2 = lesser tubercle; 3 = deltopectoral crest; 4 = posterior crest; 5 = teres tuberosity; 6 = intertubercular groove. All ×8; stereo-pairs.

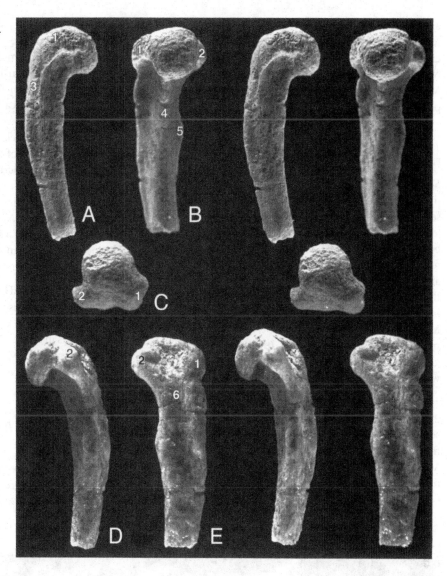

SACRUM
Figs. 18, 21

The sacrum is very long and narrow. The estimated length (measured in ventral aspect) of ZPAL MgM-I/170a (Fig. 21A, B) is 9–10 mm, and the maximum width (measured across the lateral margins of the transverse processes of the first vertebra fused to the ilia) is 6.2 mm. The lengths of the two first sacral vertebrae in this specimen are S1 2.0 mm and S2 2.7 mm. In ZPAL MgM-I/85 (Fig. 21C, D) the sacrum exposed in dorsal view is badly damaged, and its dimensions cannot be given. In ZPAL MgM-I/83 (Fig. 18), S1 is 1.9 mm long and S2 2.2 mm. In ZPAL MgM-I/83 the sacrum is compressed dorsoventrally and appears flatter than it was originally. The dorsal sacral foramina are, in dorsal view, obscured by the flattened area of the fused laminae. The first and second left dorsal foramina are preserved and seen in left lateral view. Because of deformation, the pedicles appear very low, and the intermediate sacral crest is more prominent

than it was originally. In dorsal view the fused arches form a flat triangular area that tapers caudally. The prezygapophyses are preserved only in ZPAL MgM-I/170a, exposed in ventral aspect (Fig. 21B). The transverse processes of S1 are partly preserved in ZPAL MgM-I/170a (exposed in ventral aspect) and are fused with the ilia. On both sides their anterior parts are broken, and it seems that the auricular surface was much larger (extended further cranially) than it has been preserved. A crest extends laterocaudally across the transverse process (upper arrow in Fig. 21B). The transverse processes of S2 are broken, but the facet for articulation with the right process is recognizable on the right ilium (lower arrow in Fig. 21B). We conclude that S1 and S2 articulated with the ilium as in *Kryptobaatar*.

In ventral view in ZPAL MgM-I/83 and ZPAL MgM-I/170a, the body of the first vertebra is insignificantly shorter than that of L7 (contrary to the condition in dorsal view). The transversely aligned areas of separation of the vertebrae are developed as transverse, blunt ridges. No ventral sacral foramina are preserved. There is a distinct median ventral crest on all the vertebrae, lateral to which the surfaces of the body are concave. On the preserved fragment of S4 in ZPAL MgM-I/170a, the median crest is less prominent than on the preceding vertebrae.

Pectoral girdle and forelimb
Figs. 22, 23

Scapulocoracoid. – The scapulocoracoids are preserved in three specimens (see '*Material*'). They are incomplete, represented by ventral ends. The glenoid fossa is complete on the left side of ZPAL MgM-I/111 (Fig. 22B–D). The suture between the coracoidal and scapular parts of the glenoid fossa is not discernible. The coracoid is expanded medially and laterally. As its medial margin is broken off, it is possible that the coracoid was more expanded medially than laterally, being roughly L-shaped. The acromion, preserved in ZPAL MgM-I/83 and ZPAL MgM-I/111 (Fig. 22B–D) is roughly peg-shaped as in *Nemegtbaatar*. In ZPAL MgM-I/111 the glenoid fossa and the acromion embrace the humeral head, which is preserved in articulation with the scapulocoracoid.

Humerus (Figs. 18, 22, 23). – In ZPAL MgM-I/83 (Figs. 18, 23) the proximal parts of both humeri and a poorly preserved distal part of the left humerus in articulation with the fragments of an ulna and radius are preserved. In ZPAL MgM-I/111 the proximal parts of both humeri in articulation with scapulocoracoids have been preserved (Fig. 22B–

D). The length of the humerus cannot be determined. The proximal part of the *Chulsanbaatar* humerus appears relatively gracile; it is most similar in proportions to that of *Tugrigbaatar* (Kielan-Jaworowska & Dashzeveg 1978), although in this latter genus it was possibly reconstructed too long (see 'Anatomical comparisons'). The similarity concerns the shape of the head that strongly overhangs the shaft, the position and relative sizes of the greater and lesser tubercles, the shape of the teres tuberosity and apparently the shape of the intertubercular (bicipital) groove, only the most proximal part of which has been preserved in *Tugrigbaatar*. This groove in *Chulsanbaatar* narrows distally and then widens again as in *Nemegtbaatar*, but is less deep than in *Nemegtbaatar*. Only a small part of the distal end of the left humerus preserved in ZPAL MgM-I/83 (Fig. 18, 22E, F) can be seen, and therefore it cannot be described in detail.

Radius and ulna. – The proximal part of the left ulna and radius have been preserved in articulation with the humerus in ZPAL MgM-I/83 (Figs. 18, 22E, F) and the distal end of the right radius in ZPAL MgM-I/111 (Fig. 22A). The olecranon preserved in ZPAL MgM-I/83 is relatively large. Otherwise, both specimens are very poorly preserved and do not merit description.

Pelvic girdle and hind limb

Pelvis (Figs. 18B, 21, 24, 37B). – The incomplete pelves are preserved in five specimens (see '*Material*'). The pelvis of *Chulsanbaatar* is more similar to that of *Kryptobaatar* than to

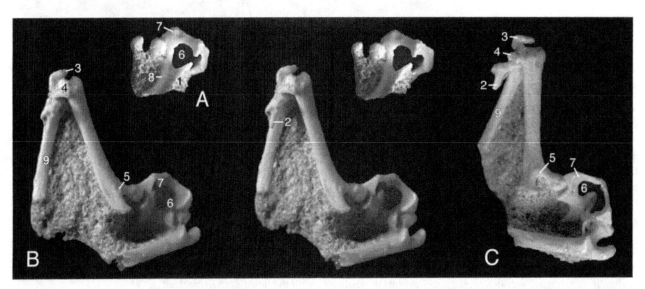

Fig. 24. ☐A–C. ?*Chulsanbaatar vulgaris* (ZPAL MgM-I/99a), Red beds of Khermeen Tsav, Khermeen Tsav II, Gobi Desert, Mongolia. Partial pubes and ischia in articulation with proximal part of left femur and nearly complete right femur in articulation with partial tibia, proximal (broken) part of fibula, patella and broken parafibula. ☐A. Pelvis in oblique lateroventral view; cranial margin up. ☐B. Same specimen, showing pelvis in dorsal view and femora arranged somewhat obliquely with respect to plane of photograph. ☐C. Same specimen with femora arranged in plane of photograph. 1 = ventral keel; 2 = broken proximal part of fibula; 3 = patella; 4 = parafibula; 5 = subtrochanteric tubercle; 6 = obturator foramen; 7 = ischium; 8 = pubis; 9 = tibia. All ×4; A, B stereo-pairs.

Fig. 25. □A–D. *Chulsanbaatar vulgaris* (ZPAL MgM-I/99b), Red beds of Khermeen Tsav, Khermeen Tsav II, Gobi Desert, Mongolia. Camera lucida drawings of left pes found in association with skeleton in Fig. 17A. Roman numerals denote the metatarsals. □A. Medial view. □B. Dorsal view. □C. Dorsolateral view. □D. Lateral view. Scale bar 5 mm.

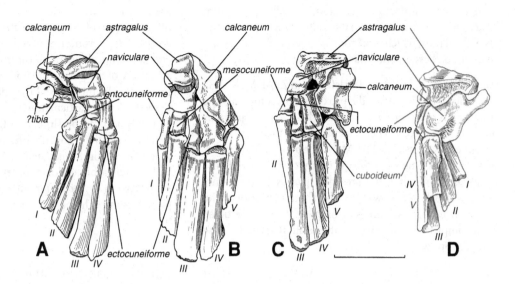

that of *Nemegtbaatar*. The estimated length of the pelvis in ZPAL MgM-I/83, based on a comparison with *Kryptobaatar*, is 20.7 mm. The ilium is very slender and elongate, 13.8 mm long in front of the acetabulum. The caudal ventral iliac spine is similar in shape to that in *Kryptobaatar*. It differs from that in *Kryptobaatar* in having the anterior extremities more strongly reflected laterally. The height of the ilium is 1.2 mm caudal to the ventral iliac spine and 2.1 mm along it. It was difficult to measure the length of the auricular surface, but it appears from the preserved parts that it was relatively shorter than in *Kryptobaatar*, the sacrum being less firmly synostosed to the pelvis. The cranial margin of the acetabulum is thickened and has a cranioventral pit. However, the prominent nodule-like thickening characteristic of *Nemegtbaatar* is absent. The body of the ilium, cranioventral to the acetabulum, is relatively smaller than in *Nemegtbaatar*, and there is no ridge dividing it into two parts. The caudal margin of the acetabulum is raised possibly higher than in *Kryptobaatar*, and there is a deep pit immediately below it, which, contrary to the condition in *Kryptobaatar* and *Nemegtbaatar*, lies within the acetabulum. The length of the acetabulum is 2.3 mm in ZPAL MgM-I/85 (Fig. 21C, D) and 1.9 mm in ZPAL MgM-I/170a (Fig. 21A, B). ZPAL MgM-I/99a (Fig. 24) represents a juvenile individual, inasmuch as the sutures between the bones of the pelvis are not synostosed. The ilium in this specimen is missing, but the preserved parts of the pubes and ischia clearly show the places of fusion with the ilium, as well as a relatively large contribution of the pubis to the acetabulum. The obturator foramen is placed ventrocaudally to it, but possibly relatively more caudally than in other genera. The obturator foramen may be measured only in ZPAL MgM-I/99a (Fig. 24), where it is 1.5 mm long and 1.3 mm high. The dorsal margin of the ischium is typically a concave arch that is oriented dorsocaudally. The ischial tuber is not preserved. The lateral side of the ischium is concave.

The ventral margin is fused with its counterpart to form a keel, the depth of which is 0.8 mm in ZPAL MgM-I/99a (Fig. 24A). The postobturator foramen is not preserved, but it appears that in ZPAL MgM-I/85 the missing part of the bone (the enlarged opening is confluent with a partly damaged obturator foramen) may have housed the postobturator foramen or notch. The pubes are not completely preserved, but it may be concluded from the preserved parts that, similar to the ischia, they were fused ventrally to form a keel.

Femur (Figs. 18, 21, 24). – The right femur in ZPAL MgM-I/83 (Fig. 18) is 18.5 mm long; in ZPAL MgM-I/99b it is still larger and more robust, while in ZPAL MgM-I/99a (juvenile individual, Fig. 24) it is 9.8 mm long. In this latter specimen the boundaries between the shaft and the epiphyses are more distinct than in other specimens, and the femoral head forms the proximal epiphysis of the femur. The femur in *Chulsanbaatar* is slender, more similar to that of *Kryptobaatar* than to that of *Nemegtbaatar*. There is some variation among studied femora, for example, in the relative size of the greater trochanter. In the smallest specimen, ZPAL MgM-I/99a (Fig. 24), and a somewhat larger one, ZPAL MgM-I/151 (both of which may not belong to *Ch. vulgaris*), the greater trochanter hardly protrudes over the femoral head proximally, while it is more prominent in larger specimens. However, even in these latter specimens it is relatively smaller than in *Kryptobaatar* and much smaller than in *Nemegtbaatar*. The gluteal crest is less prominent than in *Nemegtbaatar*. The lesser trochanter appears more prominent, more strongly protruding ventrally in *Chulsanbaatar* than in *Kryptobaatar* and *Nemegtbaatar*. The trochlea is more asymmetrically placed than in *Kryptobaatar*.

Patella. – The patella has been preserved on the left side of ZPAL MgM-I/99a (Fig. 24B, C). It sits within the knee joint below the femur, displaced laterally, arranged transversely with its convex surface oriented craniodorsally. It is 1.7 mm wide, and in comparison with the width of the distal epiphysis of the femur, which is 2.2 mm, it appears relatively wider than in small extant eutherian mammals.

Tibia (Figs. 17A, 18, 24). – The left tibia in ZPAL MgM-I/83 is 13 mm long. A part of the proximal end is missing, the distal end is complete, and no suture between the epiphysis and the shaft is seen. The proximal part of the right tibia is preserved in a juvenile ZPAL MgM-I/99a. In ZPAL MgM-I/99b (Fig. 17A) the right tibia is nearly complete, only the distal end is missing; the preserved part is 13 mm long, whereas the whole tibia was probably about 14 mm long. The proximal articular surface is incomplete. All of these tibiae are incomplete and do not show the details that would allow comparisons with *Kryptobaatar*.

Fibula (Figs. 17A, 18, 24B, C). – In ZPAL MgM-I/83 (Fig. 18) the right fibula is preserved without the proximal end. The distal part preserves the caudal tuberosity. The distal fibular end is situated caudally to the tibia. In ZPAL MgM-I/99a the proximal part of the left fibula is preserved. The head is relatively stout, and the characteristic hook-like process is preserved. In ZPAL MgM-I/99b (Fig. 17A) a nearly complete fibula, lacking only the distal extremity, is preserved in association with the tibia and femur. In all of these specimens, the fibulae are relatively more slender than in *Kryptobaatar* and *Nemegtbaatar*. In ZPAL MgM-I/99b the estimated length of the fibula is about 13.6 mm, and its width at the longitudinal midline is, in lateral view, 0.42 mm, which is 0.031 of its length, while in *Kryptobaatar* the width of the fibula is 0.059 of its length.

Parafibula. – The parafibula has been preserved only on the left side of ZPAL MgM-I/99a (Fig. 24B, C). It is a tiny, slightly cracked ossicle, similar in shape to that in *Kryptobaatar*, slightly displaced, directed dorsally, adhering with its neck to the proximal end of the fibula.

Tarsus and pes. – A left tarsus with five incomplete metatarsals has been preserved in ZPAL MgM-I/99b (Fig. 25). The hind-limb bones in this skeleton are incomplete, and their measurements are only tentative. The length of Mt III is about 3.2 mm, which is 0.24 of the estimated length of the tibia, while in *Kryptobaatar* the length of Mt III is 0.39 the length of the tibia (13.2 mm). It is thus probable that the metatarsals were relatively shorter in *Chulsanbaatar* than in *Kryptobaatar*. Because of the small size of the specimen and its poor state of preservation, we were unable to obtain good photographs. The distal bones of the tarsus have been preserved in articulation with the metatarsals. The tuber calcanei has been shifted in a plantar direction. The most medial part of the distal margin of the calcaneum fits the concavity in the cuboid. The medioproximal facet of Mt V articulates with the cuboid, while the small facet on the dorsal side of

the proximal end of Mt V apparently articulated with the distal margin of the calcaneum medial to the peroneal groove. The peroneal groove is wide, as in *Kryptobaatar*, and the peroneal tubercle is directed laterodistally. The dorsal surface of the astragalus (poorly preserved) appears relatively flat. The mediodistal rounded process with a saddle-shaped sulcus for articulation with the navicular is exposed, and it appears to be similar to that in *Kryptobaatar*. The joint between the entocuneiform and Mt I appears similarly shaped to that in *Kryptobaatar* (Fig. 7E–G). All other tarsal and metatarsal bones do not differ in shape and proportions from those in *Kryptobaatar* but, because of the poor state of preservation of the bone surface, the details of their structure cannot be given.

Family Sloanbaataridae Kielan-Jaworowska, 1974

Genus *Sloanbaatar* Kielan-Jaworowska, 1970

Sloanbaatar mirabilis Kielan-Jaworowska, 1970
(Figs. 26, 37C)

Material. – Bayn Dzak, Ruins, Djadokhta Formation, ZPAL MgM-I/20: skull, dentaries and badly damaged postcranial fragments consisting of incomplete ischia, anterior part of the left ilium, right femur, L5–L7 in articulation with damaged S1 and somewhat separated from these in the matrix, and S4 in articulation with Cd1 and Cd2.

Axial skeleton

Lumbar vertebrae. – Two lumbar vertebrae are poorly preserved and do not reveal any important details.

Sacrum. – The bones are preserved in two parts, and the estimated length of the sacrum is about 10.5 mm; the length of the last sacral vertebra is about 3.5 mm.

Pelvic girdle and hind limb

Pelvis (Figs. 26, 37C). – The preserved parts of the ilia are heavily damaged and do not differ from *Kryptobaatar*. The fragmentary ischia are strongly fused ventrally to form a keel, which is about 1.3 mm deep. The caudal part of the right ischium below the broken ischial tuber has a well preserved surface, on which we base our reconstructions of the muscular attachments of this region.

Femur. – The proximal part of the right femur is damaged, but appears to be generally similar to that of *Kryptobaatar*.

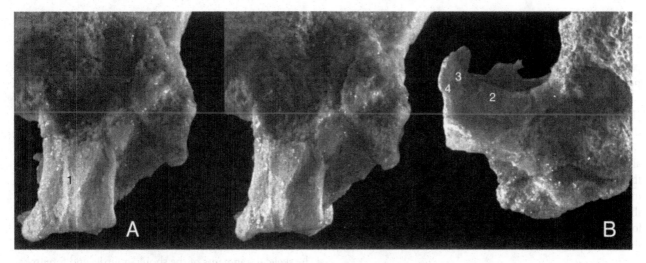

Fig. 26. □A, B. *Sloanbaatar mirabilis* (ZPAL MgM-I/20), Djadokhta Formation, Bayn Dzak, Gobi Desert, Mongolia. Broken posterior part of pelvis. □A. Ischia, ventral view. □B. Same specimen in right lateral view. 1 = ventral keel; 2–4 denote muscular origins: 2 = m. gemelli; 3 = m. biceps femoris posterior; 4 = m. semitendinosus. Both ×6; A stereo-pair.

Family Taeniolabididae Granger & Simpson, 1929

Genus *Catopsbaatar* Kielan-Jaworowska, 1994

Catopsbaatar catopsaloides (Kielan-Jaworowska, 1974)

(Fig. 27)

Synonymy. – □1974 *Djadochtatherium catopsaloides* sp. nov. – Kielan-Jaworowska, p. 40, Fig. 6, Pls. 5:9, 17:2, 18–21. □1979 *Catopsalis catopsaloides* (Kielan-Jaworowska) – Kielan-Jaworowska & Sloan, Figs. 1, 2B.

Holotype. – See Kielan-Jaworowska (1974).

Material. – Red beds of Khermeen Tsav, Khermeen Tsav II, ZPAL MgM-I/171: badly damaged broken fragments of long bones, one middle caudal vertebra and three more posterior caudal vertebrae in anatomical position. We refer this specimen to *C. catopsaloides*, as it has been found in beds yielding three skulls of this species. The postcranial fragments are of an appropriate size to be considered conspecific with the skulls described and figured by Kielan-Jaworowska (1974).

Caudal vertebrae. – The body of the middle caudal vertebra (Fig. 27A) is 10 mm long, strongly compressed laterally at its longitudinal midpoint, and slightly widening cranially and caudally; the distance between the pre- and postzyga-pophysis is 12 mm. It seems that the strong lateral compression of the vertebra is in part due to the state of preservation. The width of the body is 3.3 mm in cranial view. The spinous and transverse processes are lacking. The pre- and

postzygapophyses are very long, but the exact measurements cannot be given. In dorsal view a median ridge extends longitudinally (not preserved in the anterior one-third of the length); at the caudal end it divides into two ridges, leaving a triangular concave area in between. Lateral to the dorsal ridge there extend parallel weak ridges at the edge between the dorsal and lateral sides of the vertebra;

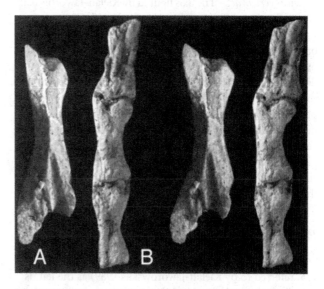

Fig. 27. □*Catopsbaatar catopsaloides* (ZPAL MgM-I/171), Red beds of Khermeen Tsav, Khermeen Tsav I, Gobi Desert, Mongolia. □A. Compressed middle caudal vertebra, dorsal view. □B. Three damaged middle caudal vertebrae, found in association with individual in A, more posterior than vertebra in A, lateral view (dorsal to the left). Fragments of the haemal arches found in association with vertebrae are not seen in the photograph. A ×4; B ×2; stereo-pairs.

these ridges continue caudally as the lateral margins of the postzygapophyses. On the ventral side of the body there are two longitudinal ridges, with a gully between them. Poorly preserved foveae for haemal arches are recognizable at the posterior end of the body.

Three caudal vertebrae from the posterior part of the tail (Fig. 27B) are strongly laterally compressed, partly because of damage. The pre- and postzygapophyses (partly broken) were shorter than in the middle caudal vertebra described above, but otherwise show a similar structure. In the last one that has been retrieved from the matrix after the photograph in Fig. 27B was taken, there is a deep longitudinal furrow at the ventral side of the body. A broken fragment of a relatively long haemal arch has been preserved between the first and the second vertebra.

Taeniolabidoid, fam. gen. et sp. indet. (Kielan-Jaworowska 1989)

Fig. 28A

Material. – Djadokhta Formation, Bayn Dzak, Main Field, ZPAL MgM-I/165: C2–C7, manubrium of the sternum, one sternebra, incomplete left scapulocoracoid, proximal part of the right humerus and fragments of ribs, described and figured by Kielan-Jaworowska (1989).

Axial skeleton

Cervical vertebrae. – The axis figured by Kielan-Jaworowska (1989, Fig. 1 and Pl. 15 and 16:1) is fused with C3, and two pairs of transverse processes are present on the fused vertebrae. The processes of C2 and C3 are very long. The caudal part of the right transverse process of C2 is broken off, but displaced, and now forms an angle with the anterior part. On the left side of this vertebra the transverse process is complete, but the suture is not preserved. The sutures cannot be discerned with any certainty on any of the transverse processes. This shows that cervical ribs were probably not present in this taxon. It also seems possible that the cervical ribs have been fused with the transverse processes because of the old age of the individual.

Pectoral girdle

Scapulocoracoid. – The scapulocoracoid of ZPAL MgM-I/165, referred to by Kielan-Jaworowska (1989) as the left, is actually from the right side. It is incomplete, similar to the *Nemegtbaatar* scapulocoracoid, from which it differs in having the ventral arcuate ridge on the blade shorter and less prominent. The cranial border is keel-shaped proximally, as in *Nemegtbaatar*, and the shallow supraspinous fossa (less obvious than in *Nemegtbaatar*) is preserved behind it.

Myological reconstructions

The areas of origin and insertion of particular muscles are generally not seen on the photographs published in this paper. In many cases, however, we were able to identify them, examining the specimens at high magnifications under Wild–Leitz binocular microscope and changing the source of the light. Nevertheless, the reconstructions of musculature that follow, as usual in fossil material, must be regarded to some extent as tentative.

Muscles of the forelimb

The skeletal fragments of the shoulder girdle and forelimbs preserved in Late Cretaceous Asian taxa described above are fragmentary. That is why we base our reconstructions of the forelimb muscles also on isolated bones from the Late Cretaceous of North America in the AMNH collection and on the humeri of ?*Lambdopsalis bulla* (referred to below as ?*Lambdopsalis*) from the Eocene of China in the IVPP collection. Because of the fragmentary nature of this material we are unable to discuss below the origin and insertion of particular muscles of the same individual or species, as we do in the case of the hind limb. As the multituberculate postcranial skeleton, at least in members of the Cimolodonta, is fairly uniform (Krause & Jenkins 1983) and differs strongly from those of other known mammals, we assume that the musculature was also homogenous in members of this suborder. Therefore we reconstruct below some muscles of cimolodont multituberculates on the basis of scars preserved on bones belonging to different genera from various geological times and geographical regions. We are fully aware of the danger of such a generalization, but otherwise the reconstruction of the muscles of the anterior part of the body would not be possible. In some cases actual reconstructions were not feasible, and we describe only the muscle scars preserved on particular bones.

Muscles of the scapulocoracoid

Fig. 28

Fragmentary scapulocoracoids have been preserved in *Kryptobaatar* (ZPAL MgM-I/41, Fig. 5B, C), *Nemegtbaatar* (ZPAL MgM-I/81, Figs. 12, 13H–J), *Chulsanbaatar* (ZPAL MgM-I/111, Fig. 22B–D) and a taeniolabidoid gen. et sp. indet. ZPAL MgM-I/165 (Kielan-Jaworowska 1989, Pl. 16:2, and Fig. 19A herein). The bone surface in *Kryptobaatar* and *Chulsanbaatar* is poorly preserved, and we base our reconstructions (Fig. 28A, C) mostly on *Nemegtbaatar* and an unidentified taeniolabidoid.

We do not know the origin of the muscles omohyoideus, trapezius p. cervicalis and omotransversarius; omohyoideus

Fig. 28. □A–C. Right scapulocoracoids. □A. Taeniolaboidoid, fam. gen. et sp. indet. ZPAL MgM-I/165, caudal view. □B. *Citellus xanthoprymnus* (ZIN 159). □C. *Nemegtbaatar gobiensis* (ZPAL MgM-I/81). Reconstructions of muscle attachments based on surface topography of bones. B, C, cranial views. Scale bars 5 mm.

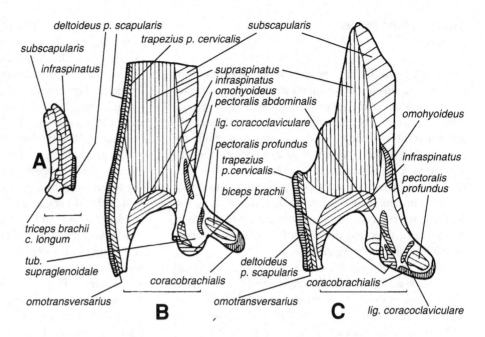

in small therian mammals usually inserts on the cranial border of the scapula, and trapezius p. cervicalis and omotransversarius insert on the scapular spine at the point of its prolongation into the acromion. At this point on ZPAL MgM-I/81 and ZPAL MgM-I/111 there extends a weak longitudinal groove, cranial to which apparently both these muscles inserted, and caudally originated m. deltoideus p. scapularis.

The presence of the incipient supraspinous fossa in *Nemegtbaatar* (see 'Osteological descriptions') indicates the existence of a relatively small m. supraspinatus, although its scars are recognizable only on the cranial surface of the spine. In extant therian mammals (Gambaryan 1960; Jouffroy 1971) it originates on the cranial surface of the spine and in the supraspinous fossa and extends onto the lateral surface of m. subscapularis; it inserts on the proximal end of the greater tubercle of the humerus. There is a depression for m. supraspinatus on the greater tubercle of the humeri of all three Mongolian taxa mentioned above and in *?Lambdopsalis* (Fig. 29).

In *Nemegtbaatar* and in unidentified taeniolabidoid scapulocoracoids we tentatively recognize four weak swellings at sites of the origin of the internal aponeuroses of the subscapularis. Three of them extend longitudinally: along the middle of the highest convexity of the medial side of the blade, along the cranial and caudal borders (the latter not preserved in *Nemegtbaatar*); the fourth is most prominent and extends as an oblique ridge in the ventral part of the blade, in an arch towards the ventral margin (number 13 in Figs. 13H and 28). If this interpretation is correct, this muscle in multituberculates was quadripennate and extended beyond the cranial border of the scapulocoracoid. Musculus subscapularis apparently inserted on the lesser tubercle, on

which there is a depression on the humeri of *Nemegtbaatar*, *Chulsanbaatar* and *?Lambdopsalis* (Figs. 13B–E, 23, 29).

As in all mammals, m. infraspinatus apparently originated from the infraspinous fossa and from the caudal surface of the scapular spine. In *Nemegtbaatar* (Figs. 13I, 28C), *Chulsanbaatar* (Fig. 22D) and the unidentified taeniolabidoid ZPAL MgM-I/165 (Fig. 28A) on the ventral end of the spine, there is a cranially situated rounded surface for the origin of this muscle, as in, e.g., *Antechinus stuarti*, *Mesocricetus raddei*, *Citellus xanthoprymnus* (Fig. 28B) and many other extant mammals. On the greater tubercle there is a depression for its insertion, well seen in the three above-mentioned multituberculate taxa. We cannot recognize in the studied multituberculates the origin or insertion of m. teres minor.

In extant mammals the origin of m. triceps brachii caput longum is well seen on the ventral end of the caudal margin of the scapula. In an unidentified taeniolabidoid, only the ventral part of this origin is visible. A part of its insertion is seen in AMNH 118505 on the proximal end of the olecranon (Fig. 14C–E; see also Figs. 32B and 33B). In *Nemegtbaatar* ZPAL MgM-I/81, on the ventral end of the coracoid there is a thickening, apparently for the origin of coracobrachialis, which extends until the horseshoe-shaped thickening for the origin of biceps brachii, at the base of the coracoid (Fig. 28C). Dorsal to the origin of m. coracobrachialis there is a small concavity, apparently for the insertion of m. pectoralis profundus, delimited ventrally by a narrow ridge for the lig. coracoclavicularis. Ventral to the origin of biceps brachii there is a small concavity interpreted as the insertion of m. pectoralis abdominalis. Part of m. pectoralis profundus inserted on the coracoid process ventral to the lig. coracoclaviculare (the other part of this muscle inserted on the humerus – see below).

Muscles of the humerus

Figs. 29–31

In ?*Lambdopsalis* (Fig. 29) the scars of the insertions of the muscles cutaneus trunci, pectoralis superficialis and pectoralis profundus are recognizable on the wide deltopectoral crest. The insertion of pectoralis profundus continues up to the greater tubercle and passes directly to the coracoid process. In this respect ?*Lambdopsalis* differs from therians (e.g., *Antechinus, Mesocricetus, Meriones* and *Citellus* [Fig. 28], studied by us) where the pectoralis profundus passes along the intertubercular groove to the lesser tubercle and then onto the coracoid process. The insertions of m. deltoideus p. clavicularis and deltoideus p. scapularis are separated in ?*Lambdopsalis* (Fig. 29).

In ?*Lambdopsalis* (Fig. 29), *Nemegtbaatar* (Fig. 13A, B) and *Chulsanbaatar* (Fig. 23A, B) the origin of m. triceps brachii caput laterale is well seen on the deltopectoral crest. Caput mediale originated in ?*Lambdopsalis* on the dorsal surface of the humerus; proximally it occupied only the medial part of this surface, and distally it extended almost across the whole width of this surface up to the epicondyles. In ?*Lambdopsalis* (Figs. 29, 30A) m. coracobrachialis apparently inserted by one head on the ventral side of the humerus reaching pronator teres. On the dorsal side of the entepicondyle (Fig. 30A) there is a distinct area for the origin of m. epitrochleoanconeus; this does not extend onto the medial surface of the entepicondyle, as characteristic of modern therians, for example, *Antechinus stuarti* and *Mesocricetus raddei* (Fig. 31). Musculus brachialis originated in ?*Lambdopsalis* (Fig. 29) on the proximal one third of the dorsal surface of the humerus, between the triceps brachii caput mediale and caput laterale. It did not reach the lesser tubercle, as in most extant therian mammals (Rinker 1954; Gambaryan 1960; Jouffroy 1971; and others).

The areas of origin of the extensor carpi radialis longus, ext. carpi radialis brevis, ext. digitorum communis and ext. digitorum lateralis are poorly preserved in ?*Lambdopsalis* and in the Mongolian taxa, but are well represented in an unidentified multituberculate from the Late Cretaceous of North America, AMNH 118267 (Figs. 14B, 30C). Scars for the origin of the extensor carpi ulnaris and supinator were not seen in all fossil taxa studied by us; however, we reconstruct them in Fig. 30 by comparison with extant taxa (Fig. 31). As in all recent mammals (Jouffroy 1971), these muscles originated one after the other on the ventral surface of the lateral ectepicondylar crest, but the boundaries between them are not recognizable and are only tentatively shown in Fig. 30C. We could not recognize the origin or insertion of m. brachioradialis. The origins of two parts of flexor digitorum profundus (situated craniodorsal and ventral to fl. digitorum sublimis), fl. carpi radialis, fl. carpi ulnaris, pronator teres, palmaris longus and lig. collaterale mediale are tentatively recognized on the entepicondyles of three taxa in Fig. 30.

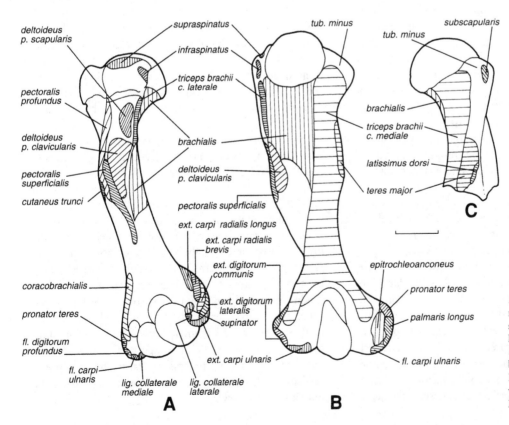

Fig. 29. Reconstruction of muscle attachments based on surface topography of bones of left humerus of ?*Lambdopsalis bulla* (IVPP V9051), Bayan Ulan beds, Early Eocene, Bayan Ulan, Inner Mongolia. □A. Ventral view. □B. Dorsal view. □C. Medial view. Scale bar 5 mm.

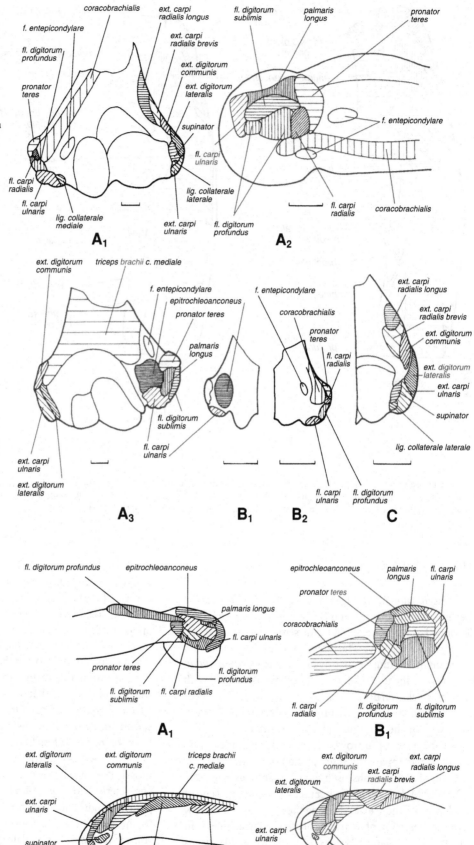

Fig. 30. Reconstruction of muscle attachments based on surface topography of distal part of left (A, C) and right (B) humeri. □A. *?Lambdopsalis bulla* (IVPP V8408). □B. *Nemegtbaatar gobiensis* (ZPAL MgM-I/81). □C. Unidentified multituberculate (AMNH 118267), Bug Creek Anthills site, Hell Creek Formation, Montana. Reconstruction of *Nemegtbaatar* muscles is hypothetical. □A₁, B₂, C. Ventral views. □A₂. Medial view. □A₃, B₁. Dorsal views. Scale bars 3 mm.

Fig. 31. Muscle attachments on distal part of right humeri in extant marsupial and eutherian mammals. □A. *Antechinus stuarti* (ZIN 1302). □B. *Mesocricetus raddei* (ZIN 660). □A₁, B₁. Medial views. □A₂, B₂. Lateral views. Scale bar 5 mm.

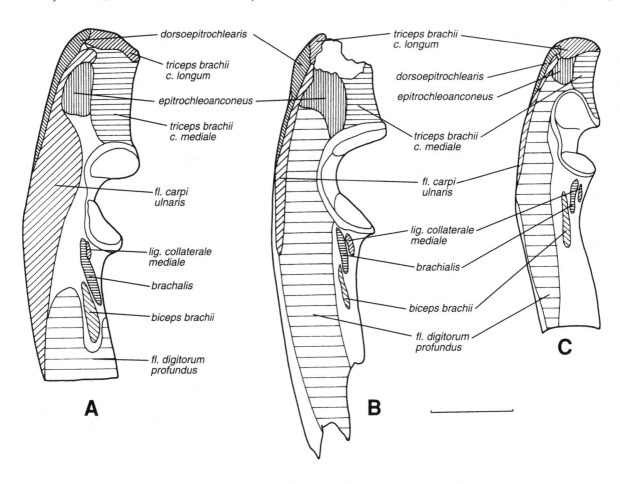

Fig. 32. Muscle attachments on proximal part of left ulnae in extant eutherian and marsupial mammals and reconstruction of muscle attachments based on surface topography of multituberculate ulna. Medial views. □A. *Mesocricetus raddei* (ZIN 660). □B. Unidentified multituberculate (AMNH 118505), Bug Creek Anthills site, Hell Creek Formation, Montana. □C. *Antechinus stuarti* (ZIN 13021). Scale bar 3 mm.

Muscles of the radius and ulna
Figs. 32, 33

In *Nemegtbaatar* ZPAL MgM-I/81, on the proximal part of the radius the tuberosity for the insertion of biceps brachii is developed as a groove, distal to the semilunar articular circumference (Fig. 14A). The insertions of biceps brachii and brachialis on the ulna are poorly preserved in *Nemegtbaatar*. However, on the ulna AMNH 118505, the insertion of both muscles is well seen distal to the coronoid process (Figs. 14C, D, 32B). In many therian mammals m. biceps brachii does not insert on the ulna but on a tuberosity on the radius (Gambaryan 1960; Jouffroy 1971). Lig. collaterale mediale in *Nemegtbaatar* apparently inserted cranioproximally to the insertion of m. brachialis, as in therian mammals (Fig. 32).

On the AMNH 118505 ulna, caudal to the proximal end of the olecranon, there extends along the lateral side a narrow furrow for insertion of m. triceps brachii caput laterale, as in modern therian mammals (Figs. 14C, 33). Musculus triceps brachii c. mediale inserted on the cranial, craniomedial and craniolateral surfaces of the olecranon. In this specimen, insertion of m. epitrochleoanconeus is seen on the medial surface of the olecranon (Fig. 14E, 32B). On the caudal side of the medial margin there is a furrow apparently for the aponeurosis of m. dorsoepitrochlearis. Between it and the insertion of m. epitrochleoanconeus extends the origin of flexor carpi ulnaris. Most of the medial side was occupied by the origin of fl. digitorum profundus, the shape of which is more reminiscent of that of *Antechinus stuarti* than of *Mesocricetus brandti* (Fig. 32). On the lateral surface of the same specimen, between the insertion of the medial and lateral heads of m. triceps brachii, there is a small tuberosity for the origin of extensor carpi ulnaris. The origin of ext. carpi ulnaris does not reach the origin of ext. pollicis longus (as characteristic for, e.g., *Antechinus stuarti*). The origin of ext. pollicis longus and abductor pollicis longus is more similar to the condition in *Mesocricetus brandti* than to that in *Antechinus stuarti*; lig. collaterale laterale inserted cranial to abd. pollicis longus (Fig. 33, see also Fig. 14E).

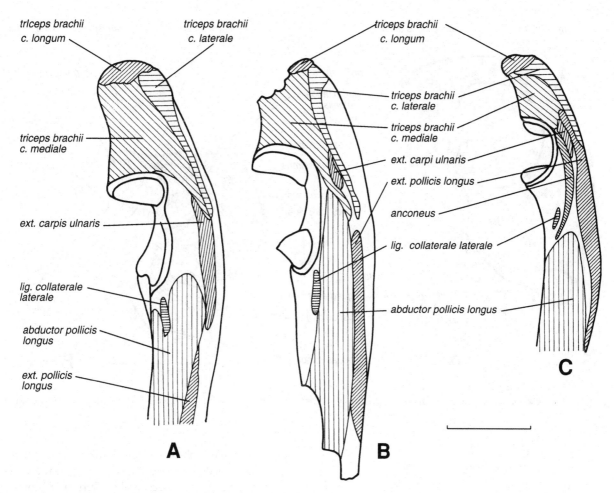

Fig. 33. Same specimens as in Fig. 32 in lateral views. Scale bar 3 mm.

Muscles of the axial skeleton

Figs. 34–41

A comparison with extant small therian mammals shows that the structure of the lumbar vertebrae depends upon the development of the musculi erector spinae (longissimus dorsi, semispinalis dorsi and iliocostalis dorsi). The lengths of the transverse processes are correlated with the mass of the epaxial musculature (Fig. 47). The interrelationships between m. longissimus dorsi and semispinalis dorsi depend on the distance between the prezygapophysis and the spinous process (Figs. 34, 36). On the ventral side of the lumbar vertebrae there are, in therians, three muscles: psoas minor, psoas major and quadratus lumborum (Fig. 34B). There is a high ventral crest on the ventral surface of L2 in *Nemegtbaatar*. In extant mammals a prominent ventral crest (referred to as the ventral spinous process) occurs only among leporids, in which m. psoas major is strongly developed.

In extant mammals m. quadratus lumborum arises by tendons from the bodies of the last two or three thoracic vertebrae and has a fleshy origin on the bodies of the first four lumbar vertebrae (Jouffroy & Lessertisseur 1968). Its tendons of insertion are placed at the craniolateral corners of the transverse processes of the lumbars and on the first sacral vertebra, close to the articulation with the ilium. Between the dorsal surface of the muscular origin and the ventral surface of the tendons of insertion there are multipennate muscle fibers. In *Nemegtbaatar*, on the transverse processes on the right side of L4 and L7 and the left side of L5, there is a distinct prominence. We speculate that m. quadratus lumborum inserted by internal aponeurosis on the ventral side of the craniolateral corners of the transverse processes and was very strong (Figs. 34A, 35, 39B).

The dorsal part of the epaxial musculature is complex in extant therian mammals (Slijper 1946; Jouffroy & Lessertisseur 1968). Musculus interspinalis consists of right and left bundles of fibers that extend between the spinous processes of two consecutive vertebrae. The right and left bundles (dorsal view) are separated by a vertical tendon (Fig. 36A). The presence of the medial and lateral ridges on the caudal surface of the spinous process in *Nemegtbaatar* (ZPAL

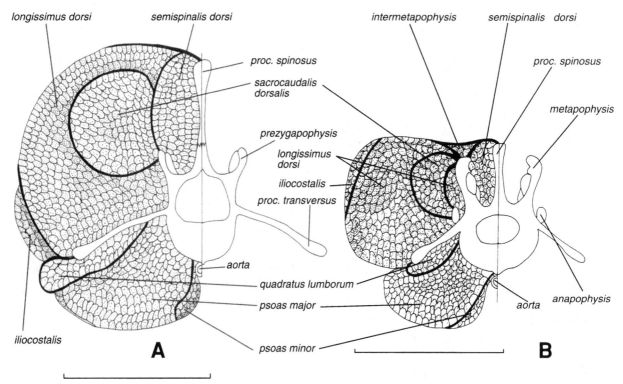

Fig. 34. □A. *Nemegtbaatar gobiensis.* Reconstruction of transverse section of vertebral column at level of L5; □B. Same region in *Meriones blackleri.* Scale bar 10 mm.

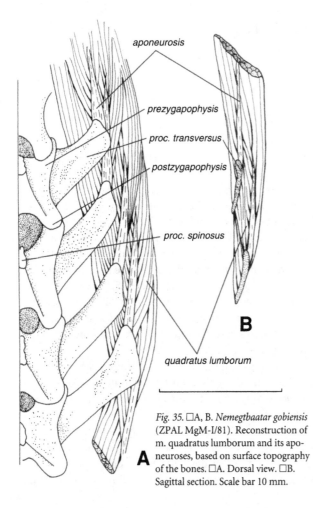

Fig. 35. □A, B. *Nemegtbaatar gobiensis* (ZPAL MgM-I/81). Reconstruction of m. quadratus lumborum and its aponeuroses, based on surface topography of the bones. □A. Dorsal view. □B. Sagittal section. Scale bar 10 mm.

MgM-I/81), seen only on L3 and L4 (the remaining spinous processes are badly damaged), indicates that m. interspinalis (inserted between the ridges) and the vertical tendon were well developed (Fig. 36B, see also Fig. 11).

In *Nemegtbaatar,* m. intertransversarius was apparently weakly developed, as the insertion area for this muscle, on the cranial surface of the transverse process is very small (Figs. 11, 36B).

Musculus intermetapophysis (intermamillaris) in rodents studied by us (Figs. 34B, 36A) arises on the cranial side of the metapophyses of S1 and all the lumbars, and is inserted on the caudal side of these metapophyses. In addition, on L6 to L1, between the metapophyses and anapophyses of every other vertebra (e.g., metapophysis of L6 – anapophysis of L4; metapophysis of L5 – anapophysis L3), there extends another branch of this muscle. In mammals that have this latter muscle developed (for example in all extant taxa cited in Table 2), the metapophyses show, in dorsal view, both medial and lateral deviation. Musculus intermetapophysis was apparently not present in *Nemegtbaatar* (Figs. 34A, 36B). The metapophyses are absent in *Nemegtbaatar,* but the prezygapophyses (which on the right side of L7, left L6 and L4 have dorsal surfaces well preserved), do not show the medial deviation characteristic of this muscle (Fig. 36B).

We conclude that the tail in *Nemegtbaatar* (not preserved) was very long. In rodents and some insectivores (for example *Ellobius, Talpa*) with a short tail, m. sacrocaudalis dorsalis does not reach the lumbar vertebrae. In mammals with a medium-size tail, this muscle arises from the metapophysis

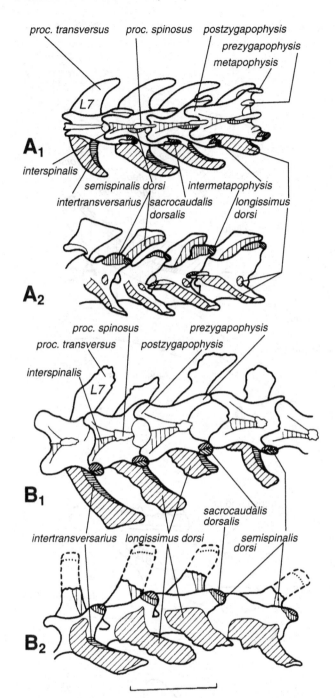

Fig. 36. Reconstruction of muscle attachments based on surface topography of the last lumbar vertebrae. Cranial to the right. □A. *Meriones tamariscinus* (ZIN 428), lateral view. □B. *Nemegtbaatar gobiensis* (ZPAL MgM-I/81). A₁, B₁, dorsal view. A₂, B₂, lateral view. Dashed and dotted lines in B₂ indicate the two tentative reconstructions of the lengths of the spinous processes. Reconstructions based on originally preserved more complete spinous processes (now broken) in ZAPL MgM-I/81 and on relatively long (partly broken) spinous processes of L6 and L7 in *Kryptobaatar* (Fig. 2). Scale bar 10 mm.

of the two posterior lumbars (*Microtus, Microgale*). In mammals with a long tail, m. sacrocaudalis dorsalis arises from the metapophyses of all the lumbars (Figs. 34B, 36A). In ricochetal mammals with a very long tail (*Allactaga, Rhyncho-*

cyon) this muscle also reaches the last thoracic vertebrae. In *Nemegtbaatar* this muscle originated from the caudal margins of the prezygapophyses of the last four lumbar vertebrae (the upper arrow in Fig. 11A). As the anterior lumbars are damaged, it cannot be stated whether sacrocaudalis dorsalis continued more cranially; it is tentatively reconstructed in Figs. 34A, 36B, 39, 40B, 41A.

The ventral side of the sacrum has not been exposed in *Kryptobaatar* and *Nemegtbaatar* and could be examined only in *Chulsanbaatar* (Fig. 21B). The laterocaudally directed ridge on the transverse process of S1 apparently formed the cranial boundary of the insertion of m. sacrocaudalis ventralis. If this were true the muscle would not extend onto the lumbars (Fig. 39, 40, 41A).

Musculus longissimus dorsi is the largest of the epaxial musculature in small extant mammals (Rinker 1954; Jouffroy & Lessertisseur 1968). It arises from the medial and cranial surfaces of the ilium (two upper left arrows in Fig. 16E) and inserts on the dorsal surface of all the transverse processes of the lumbar vertebrae and also on the anapophyses. In *Nemegtbaatar* there are no anapophyses, but on the pedicles, below the medial sacral crest, there is a weak longitudinal ridge, convex upwards. This ridge corresponds to the origin of m. longissimus dorsi, which extended from the ilium cranioventrally and then in an arch craniodorsally, below m. sacrocaudalis dorsalis (Figs. 34A, 37A, 39, 40, 41A, see also Fig. 11). The part of this muscle that in modern mammals arises from the spinous processes and inserts on the thoracic vertebrae could not be reconstructed, as the spinous processes are damaged, and the thoracic vertebrae are not preserved. Musculus semispinalis dorsi originated in *Nemegtbaatar* from the cranial end of the prezygapophyses that are medially deviated (the lower arrow in Fig. 11A and Figs. 34A, 36B, 39, 40).

Muscles of the pelvic girdle and hind limb

Our reconstructions of the pelvic and hind-limb musculature of Mongolian taxa differ in details from those presented by Simpson & Elftman (1928) for ?*Eucosmodon*. We follow Gambaryan's (1974) division of the hind-limb musculature in therian mammals, which provides the basis for functional analysis.

Muscles of the hip joint
Figs. 37–44

Gambaryan (1974) recognized the following groups of muscles of the hip joint: flexors (rectus femoris, iliopsoas); three groups of extensors: gluteal group (gluteus superficialis, gluteus medius, gluteus profundus and piriformis), short postfemoral muscles (pectineus, add. longus, add. brevis, add. magnus, praesemimebranosus and quadratus

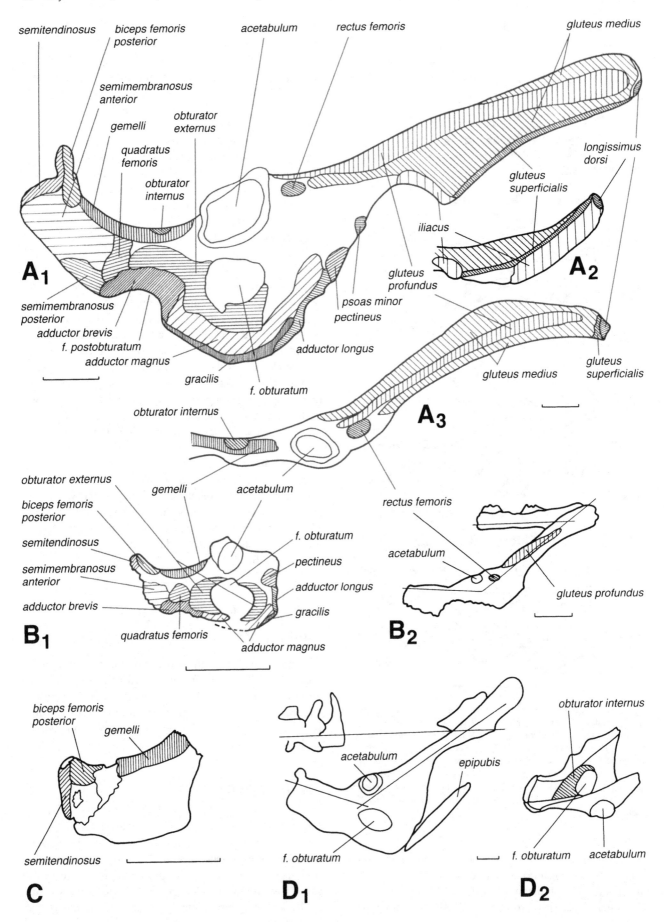

femoris); short muscles of the trochanteric fossa (obturator externus, obturator internus, gemelli cranialis and gemelli caudalis); long postfemoral muscles (biceps femoris, gracilis, semitendinosus, semimembranosus and tenuissimus – the latter not always present). Musculus psoas minor, which we also describe below, does not belong to the above defined groups.

In extant mammals (for example in those cited in Table 2) m. psoas minor arises from the middle part of the ventral surface of the last thoracic and first lumbar vertebrae; it inserts on the psoas tubercle of the pubis (Fig. 38). We were unable to find its origin in studied multituberculates. However, in *Nemegtbaatar* we found its insertion at the cranial end of the transverse ridge of the dorsal part of the pubis (Figs. 16F, 37A$_1$), where at the cranial part of this ridge there is a small tuberosity with a small depression dorsal to it. Apparently the insertion of this muscle was both fleshy and tendinous. The dorsal concave side of the described structure corresponds to the fleshy area of attachment, while the cranioventral tuberosity corresponds to the tendinous attachment. In extant mammals the insertion of this muscle is tendinous; the apparent fleshy–tendinous insertion in multituberculates possibly represents a more primitive stage.

Flexors of the hip joint (Fig. 37–42). – Musculus rectus femoris in all of the studied multituberculates originated from a shallow, round depression situated cranial to the small node with a pit that lies directly in front of the margin of the acetabulum (Figs. 2, 16F, 37A, B, 39A, 40B, 41B). In most extant mammals this area of attachment is developed as a tuberosity (Fig. 38). The presence of a depression for the origin of rectus femoris in multituberculates suggets that its origin was fleshy.

We were unable to recognize the origin of m. psoas major in studied multituberculates. However, the ventral crest preserved on the body of L2 in *Nemegtbaatar* indicates a strong development of this muscle. This conclusion is also supported by an apparent area of its insertion on the lesser trochanter, which in all multituberculates is very prominent (Figs. 2B, 16A, C, 21A). Psoas major probably inserted on the medial side of the lesser trochanter (Fig. 42A–C). Musculus iliacus in multituberculates, because of the dorsoventral iliosacral contact, originated from the ventrolateral surface of the ilium (Fig. 37A). In this respect multituberculates differ from small extant therian mammals, e.g., *Antechinus* and *Meriones*, in which iliacus originates from the ventral part of the lateral side of the ilium (Fig. 38). Iliacus inserted in multituberculates on the lesser trochanter, apparently by a common tendon with psoas major (as in extant mammals), and is not shown in Fig. 42.

Gluteal group (Figs. 37–42). – In *Nemegtbaatar* the area of origin of m. gluteus superficialis extended continuously along the ventral margin of the iliac wing from the caudal ventral iliac spine to the cranial end of the ilium (Fig. 37A). The origin of this muscle on the wing of the ilium shows that it was not divided as yet into tensor fasciae latae and gluteus superficialis. Therefore we reconstruct these two muscles as one muscle (Fig. 40A). In extant therian mammals, as a rule, gluteus superficialis arises also from the gluteal fascia, which extends between the spinous processes of the last lumbar and first caudal vertebrae (Fig. 38). Apparently this was also the case in the studied multituberculates, as indicated by the inferred site of insertion of this muscle, which extends continuously distally along the lateral edge of the femur; the distal boundary of the insertion area could not be recognized (Figs. 16A, B, 40A, 41C, 42C, D). Musculus sartorius, recognized by Simpson & Elftman (1928) in ?*Eucosmodon*, was not developed in multituberculate taxa studied by us.

Musculus gluteus profundus in studied multituberculates was surrounded dorsally, laterally and ventrally by gluteus medius (Figs. 37A$_{1-3}$, B$_2$, 39, 40, 41C). The area of origin of gluteus profundus is clearly seen on the lateral surface of the ilium (right arrows in Fig. 16D); dorsal and ventral to it there is an apparent area of attachment of gluteus medius up to the boundary with the attachment of gluteus superficialis (Fig. 37A); the dorsal origin of gluteus medius extends along the longitudinal ridge on the dorsal side of the ilium ((left arrows in Fig. 16D and right arrows in Fig. 16E). In *Nemegtbaatar* the area of attachment of gluteus profundus extended caudally as far as the area of origin of rectus femoris, and gluteus medius (in dorsal view) extended up to the middle of the medial side of the thickened margin of the acetabulum (Fig. 37A). In many small extant mammals, in contrast, gluteus profundus extends more caudally than gluteus medius (Fig. 38). In all studied multituberculates it was impossible to establish the origin of m. piriformis, which occurs in small extant eutherian mammals (Fig. 38A). The insertions of gluteus medius, gluteus profundus and piriformis are clearly seen on the greater trochanter (especially in *Nemegtbaatar*) and along the lateral edge of the femur (Figs. 16B, 42C, D). Musculus gluteus profundus inserted on the most medial part of the rugose area of the greater trochanter; proximal to it, on the top of the trochanter, inserted the piriformis; the gluteus medius extended from the top of the trochanter along the lateral edge of the femur, up to the contact with gluteus superficialis. The boundary between the insertion of these two muscles is not clear.

Fig. 37. Reconstruction of muscle attachments based on surface topography of multituberculate pelves. □A$_1$. *Nemegtbaatar gobiensis* (ZPAL MgM-I/81). Lateral view of pelvis. □A$_2$. Latroventral view of iliac wing of A$_1$. □A$_3$. Dorsal view of anterior part of A$_1$. □B$_1$. *Chulsanbaatar vulgaris* (ZPAL MgM-I/99a). Caudal part of pelvis, lateral view. □B$_2$. *Chulsanbaatar vulgaris* (ZPAL MgM-I/85). Almost entire pelvis, lateral view. □C. *Sloanbaatar mirabilis* (ZPAL MgM-I/20). Lateral view of the posterior part of the pelvis. □D$_1$. *Kryptobaatar dashzevegi* (ZPAL MgM-I/41). Lateral view of pelvis (mirror image of left side) and caudal vertebrae. □D$_2$. Dorsal view of posterior part of D$_1$, slightly inclined to left side (mirror image). Straight lines in B$_2$ and D$_1$ indicate long axes of ilium, ischium and sacrum. Iliosacral angle is between long axes of ilium and sacrum. Scale bars 3 mm. (See Fig. 38 for comparison with extant marsupial and eutherian mammals.)

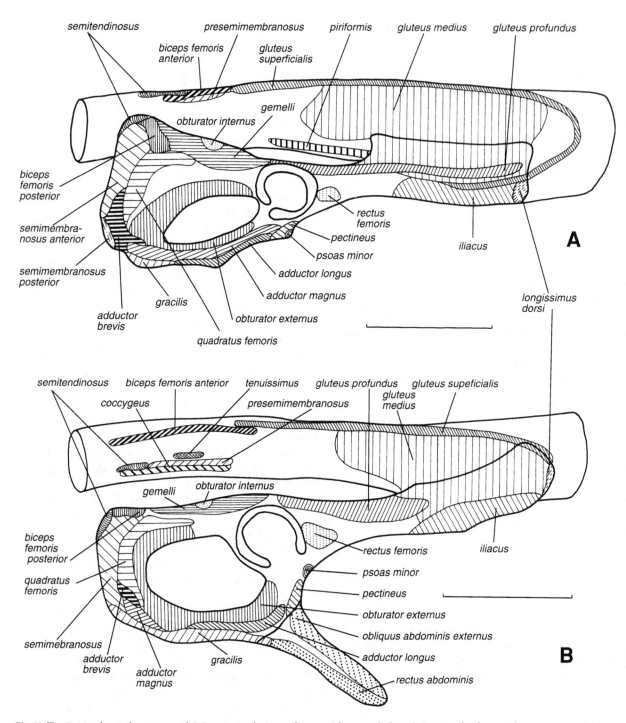

Fig. 38. □A, B. Muscle attachments on pelvis in extant eutherian and marsupial mammals, lateral view. Last lumbar vertebrae, sacrum and first caudal vertebrae diagrammatically illustrated as tube. Musculi erector spinae and m. sacrocaudalis dorsalis, on which fasciae of the pelvic muscles originate, are not shown in this figure. □A. *Meriones blackleri*. □B. *Antechinus stuarti*. Scale bars 5 mm.

Short postfemoral muscles (Figs. 37–42). – In multitubercu-lates m. pectineus originated from a rounded, relatively large area of the pubis, below the transverse ridge that extends cranially from the lower margin of the acetabulum. This area is comparatively larger in *Chulsanbaatar* than in other gen-era (Fig. 37B$_1$). This muscle inserted along the proximal half of the medial edge of the femur and was relatively strong in multituberculates (Fig. 42A, B).

Musculus adductor longus in *Nemegtbaatar* (Figs. 37A$_1$, 41A, B) originated from the cranioventral margin of the pubis, immediately ventral to m. pectineus, and surrounded the area of articulation with the epipubic bone. It covered the cranial part of the origin area of m. gracilis, the craniodorsal part of which extended between add. longus and add. magnus. In *Chulsanbaatar*, gracilis apparently did not ex-tend between add. longus and add. magnus (Fig. 37B$_1$).

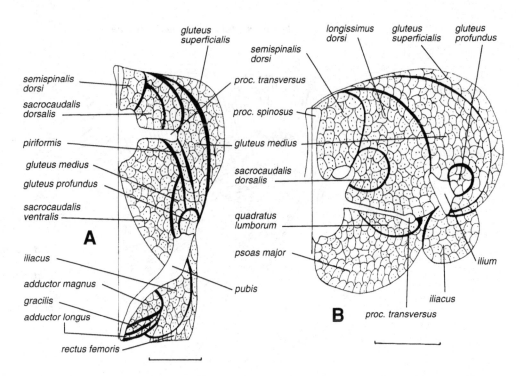

*Fig. 39. Nemegtbaatar
gobiensis* (ZPAL MgM-I/81).
□A. Transverse section of
pubic region in front of
acetabulum, at level of S3.
□B. Transverse section of
ilium at level of L7. Scale
bars 5 mm.

Adductor longus inserted on the ventral side of the femur, immediately lateral to the insertion of pectineus (Fig. 42A, B). Adductor magnus in *Nemegtbaatar* originated from the ventral part of the lateral side of the pubis and ischium, immediately dorsal to gracilis and add. longus; add. magnus reached with its cranial end the level of pectineus (Fig. 37A₁). In *Chulsanbaatar* it did not extend so far dorsally (Fig. 37B₁). Adductor magnus in *Nemegtbaatar* inserted on the ventral side of the femur, occupying a large area (about two thirds of the ventral surface of the femur) between the lateral and medial borders (Fig. 42A, B). Adductor brevis in *Nemegtbaatar* and *Chulsanbaatar* (Figs. 37A₁, B₁) extended from the ventral part of the lateral side of the ischium, immediately to the rear of add. magnus, surrounding the postobturator foramen (or notch), characteristic of multituberculates. Adductor brevis inserted on the lateral margin of the femur, distal to the lesser trochanter, as far as midway along the femur to a point between the insertions of add. magnus and gluteus superficialis (Fig. 42B, C).

Musculus quadratus femoris in *Nemegtbaatar* (Fig. 37A₁) originated from the ischium, among the following muscles: semimembranosus anterior, semimembranosus posterior, add. brevis, obturator externus and gemelli. It inserted on the medial side of the lesser trochanter of the femur (Fig. 42B, C). The insertion area on the trochanter is surrounded by a weak ridge, on which there are small transverse ridges, apparently indicating the multipennate structure of this muscle.

Musculus praesemimembranosus (caudofemoralis) in small extant mammals, as a rule, originates from the spinous or transverse processes of the sacrum (Fig. 38). We were unable to identify its site of origin in any of the multi-

tuberculate specimens described here. However, on the femur of *Nemegtbaatar*, we infer that the area of insertion is situated medial to the origin of gastrocnemius medialis (Figs. 17D, 42A, B). The caudofemoralis identified by Simpson & Elftman (1928) in ?*Eucosmodon* is not homologous to the muscle described by us under the same name. Musculus caudofemoralis of Simpson & Elftman corresponds to the caudal part of gluteus superficialis (= femorococcygeus of Dobson 1882–1890; Jouffroy 1971 and others).

See also Figs. 40, 41 for the reconstruction of the above-discussed muscles and Fig. 38 for comparisons with extant mammals.

Long postfemoral muscles (Figs. 37–41, 43). – Musculus gracilis originated in studied multituberculates from the area in front of the postobturator notch on the ventral keel of the pelvis and extended along the ventral margin of the pubis. In *Chulsanbaatar* it reached the caudal margin of adductor longus, whereas in *Nemegtbaatar* it entered between the bellies of add. longus and add. magnus (Fig. 37A₁, B₁). In most extant mammals, as a rule, gracilis is the most medial muscle of the thigh group (ventral in Fig. 38) and is not covered by other muscles. However, in *Allactaga* and *Sorex*, for example, we found gracilis in a position similar to that in *Nemegtbaatar* (Fig. 41A, B), i.e. covered laterally by adductor longus. We conclude, therefore, that in multituberculates gracilis did not extend onto the epipubic bone. This is unlike the condition in *Ornithorhynchus* and *Tachyglossus*, where it does extend onto the epipubic bone (Gregory & Camp 1918). In *Antechinus* (Fig. 38B) it extends only to the proximal part of the epipubic bone. In *Nemegtbaatar*, gracilis

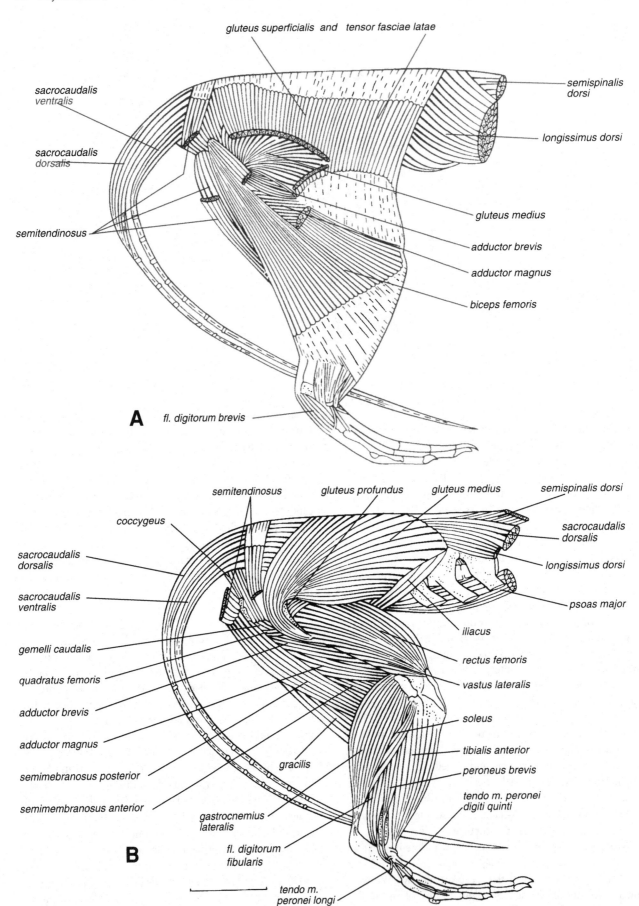

gluteus superficialis and tensor fasciae latae

semispinalis dorsi

longissimus dorsi

sacrocaudalis ventralis

sacrocaudalis dorsalis

semitendinosus

gluteus medius

adductor brevis

adductor magnus

biceps femoris

A *fl. digitorum brevis*

semitendinosus *gluteus profundus* *gluteus medius* *semispinalis dorsi*

coccygeus

sacrocaudalis dorsalis

longissimus dorsi

sacrocaudalis dorsalis

sacrocaudalis ventralis

psoas major

gemelli caudalis

iliacus

quadratus femoris

rectus femoris

adductor brevis

vastus lateralis

adductor magnus

soleus

semimebranosus posterior

tibialis anterior

peroneus brevis

semimembranosus anterior

gracilis

tendo m. peronei digiti quinti

gastrocnemius lateralis

B

fl. digitorum fibularis

tendo m. peronei longi

inserted on the medial side of the tibia in the long groove that extends along two-thirds of the tibial length and continued more proximally than, for example, in *Antechinus*.

Musculus biceps femoris posterior (we were unable to reconstruct biceps femoris anterior) and semitendinosus are only tentatively reconstructed here for *Nemegtbaatar*. We reconstruct them as originating from the ischial tuber: biceps femoris posterior cranially, semitendinosus caudally (Fig, 37A$_1$). The reason for such a reconstruction is the presence of well-preserved areas for the ventral part of the attachment of these muscles in *Sloanbaatar* (Figs. 26B, 37C) and, less clearly, in *Chulsanbaatar* (Fig. 37B$_1$). It appears that in multituberculates semitendinosus also had a vertebral head on the sacral and caudal vertebrae (Fig. 40). In most extant mammals this muscle has two heads (one ischial and one vertebral, Fig. 38). Verheyen (1961) believed that absence of the vertebral head is related to swimming habits, as he noted in, e.g., *Potamogale* and *Micropotamogale*. There is no reason to postulate that the vertebral head disappeared in studied multituberculates, although Simpson & Elftman (1928) believed that it was not present in ?*Eucosmodon*. Semitendinosus apparently inserted in the same groove as gracilis, on the medial side of the tibia. In *Nemegtbaatar* the distal end of this groove widens, apparently for insertion of this muscle (Figs. 41A, B, 43C, D). In extant mammals semitendinosus and gracilis insert in the wider part of this groove. The site of insertion of biceps femoris posterior is not visible on the bones of extant mammals and has not been recognized in studied multituberculates. Possibly, as in extant mammals (e.g., those cited in Table 2), this muscle inserted in *Nemegtbaatar* on the ventral (cranial) side of the tibia, continued on the fascia cruris and possibly on the common calcaneal tendon (Fig. 40A).

Musculus semimembranosus anterior arises from a large area on the caudal part of the ischium in a variety of recent therians (Fig. 38). From the size of this area in multituberculates one can infer that this muscle was especially strongly developed (Fig. 37A$_1$, B$_1$). It inserted on the well-defined area on the medial side of the proximal end of the tibia (arrow in Fig. 17I). As the tubercle for the lig. collaterale mediale is situated distally, and its central area is elongate, we conclude that semimembranosus anterior passed under this ligament (Figs. 41B, 43C, D). Musculus semimembranosus posterior originated from the ischium, ventral to semimembranosus anterior, and was surrounded cranially by add. brevis and dorsally by quadratus femoris. It inserted on the medial side of the tibia, ventral (cranial) to the insertion of popliteus and distal to the lig. collaterale mediale (Fig. 43C, D).

Fig. 40. Nemegtbaatar gobiensis. Reconstruction of muscles of right hind limb, lateral views. □A. Superficial layer. Musculus gluteus superficialis and tensor fasciae latae reconstructed as one muscle (see p. 49). □B. Deep layer. Scale bar 10 mm.

Short muscles of the trochanteric fossa (Figs. 37, 38, 41A, C, 42B). – Musculus gemelli originated from the dorsal side of the ischium. At the dorsal margin of the ischium there is in *Nemegtbaatar* a small oval concavity for accommodation of the tendon of m. obturator internus (Fig. 37A$_{1,3}$). In general this tendon in extant mammals divides the attachment of gemelli into gemelli cranialis and gemelli caudalis. In studied multituberculates this division was not complete. The area of the origin of obturator internus is clearly seen in *Kryptobaatar* (Fig. 37D$_2$), and on this basis we reconstruct it in *Nemegtbaatar* (Fig. 41A). It originated from the medial surface of the ischium and ran along the border of the obturator foramen dorsocaudally. Its tendon embraced the ischium dorsally and partially extended onto the femur. Musculus obturator externus originated from a large area on the lateral side of the pubis and ischium, ventrocaudally to the obturator foramen (Figs. 37A$_1$, B$_1$, 41C). In extant mammals the short muscles of the trochanteric fossa insert in this fossa. As a rule, obturator internus is situated most proximally; distally to it insert both gemelli and most distally obturator externus (Fig. 38). Apparently, this musculature inserted in studied multituberculates in a similar way.

Muscles of the knee joint
Figs. 37–42, 44

Extensors of the knee joint. – Musculus quadriceps femoris in mammals (Fig. 38) consists of four heads, three of which (vastus lateralis, vastus medialis and vastus intermedius) originate from the femur and the fourth (rectus femoris) from the ilium. The latter has been described above as one of the flexors of the hip joint. Musculus rectus femoris and the three heads of vastus work together as extensors of the knee. In *Nemegtbaatar* the area of origin of vastus lateralis is well preserved. This muscle originated as an aponeurosis from the crenulated ridge on the trochanter major, distal to the insertion of gluteus profundus, and extended medially to the subtrochanteric tubercle. Distal to the subtrochanteric tubercle, m. vastus lateralis fits tightly with vastus medialis. The area of origin of these two muscles occupied the proximal one-third of the dorsal surface of the femur. The distal one-third of this surface was occupied by the area of attachment of vastus intermedius (Figs. 41A, B, 42A, C, D). All four heads apparently inserted on the patella, which, in the studied multituberculates, has been preserved only in *Chulsanbaatar* (Fig. 24B, C).

Flexors of the knee joint. – Musculus popliteus apparently originated in *Nemegtbaatar* (Figs. 17C, 41B, 42C, 43C, D), from the ventral part of the lateral epicondyle of the femur (as in modern therian mammals). Its insertion on the mediocaudal surface of the tibia was triangular in shape, distally tapered and occupied the proximal one third of the tibial surface.

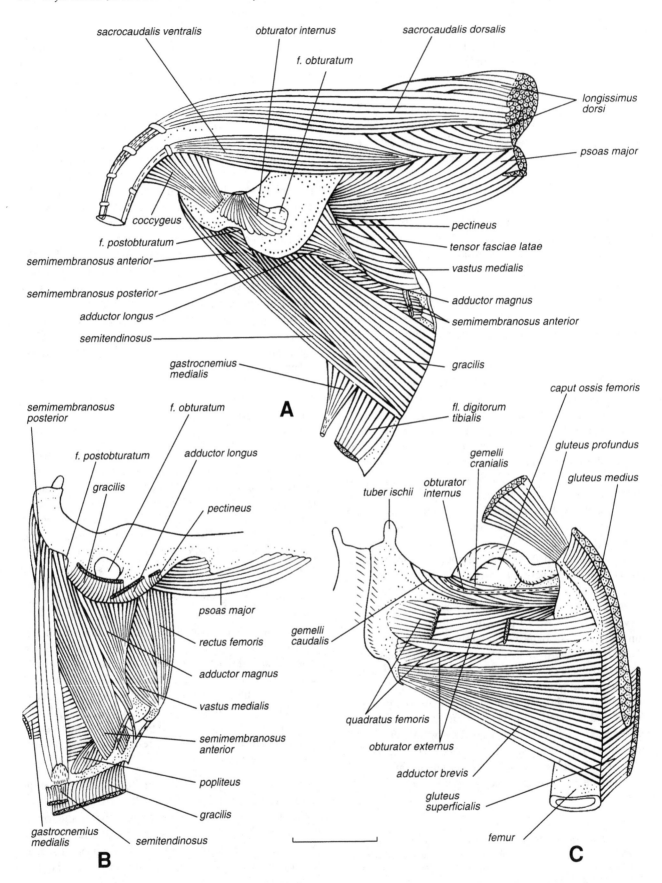

Fig. 41. Nemegtbaatar gobiensis. □A, B. Mirror images of reconstructions of muscles of right hind limb in medial views. □A. Superficial layer. □B. Deep layer. □C. Reconstruction of muscles of right hind limb in ventrocaudal view. Scale bar 10 mm.

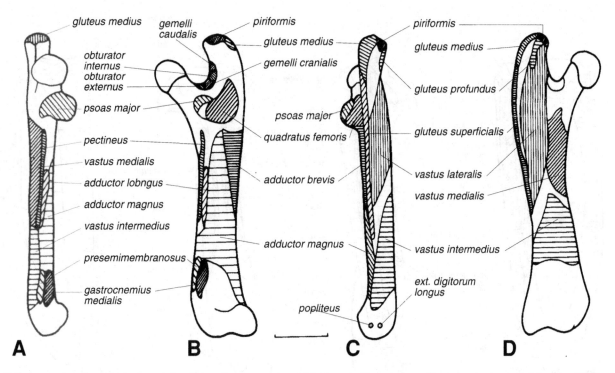

Fig. 42. Reconstruction of muscle attachments based on surface topography of right femur in *Nemegtbaatar gobiensis*, proximal two thirds based on ZPAL MgM-I/81, distal part on ZPAL MgM-I/110 (reversed). □A. Medial view. □B. Ventral view. □C. Lateral view. □D. Dorsal view. Scale bar 5 mm.

Muscles of the tarsal joint
Figs. 40–44

In *Nemegtbaatar* m. gastrocnemius medialis originated, as in modern therians (Jouffroy 1971), from the medioventral surface of the femur, in a shallow depression proximal to the medial condyle (Figs. 17E, 42A, B). The area of origin of this muscle has not been exposed in *Kryptobaatar*.

In all eutherians m. gastrocnemius lateralis originates from the femur (Jouffroy 1971). In some extant marsupials, e.g., *Antechinus stuarti* (Fig. 44B), gastrocnemius lateralis originates not from the femur but from the parafibula. The parafibula has been preserved in *Kryptobaatar* (Figs. 2, 3A, B, 44A), *Chulsanbaatar* (Fig. 24B, C) and an uncertain fragment in *Nemegtbaatar* (Fig. 17F, H, I). The area of the origin of gastrocnemius lateralis is seen on the parafibula in *Kryptobaatar*, where there is a distinct concavity for its origin (less obvious in *Chulsanbaatar*). As in *A. stuarti*, in studied multituberculates the parafibula was probably connected by a strong ligament with the patella, which has been preserved only in *Chulsanbaatar* (Fig. 24B, C). Gastrocnemius lateralis, which originates on the parafibula in *A. stuarti* (on the dorsal side of which inserts lig. parafibularo–patellare) acts as extensor of the knee joint and assists in the extension of the tarsal joint (referred to as flexion in human anatomy, see Davies & Davies 1962). We conclude that in studied multituberculates gastrocnemius lateralis acted as in *A. stuarti* (Fig. 44B).

In *Nemegtbaatar* (Figs. 17F, H, 43A, B) and, less obvious, in *Kryptobaatar* (Fig. 2) on the laterocaudal part of the

proximal end of the fibula there is a surface evidently corresponding to the origin of m. soleus. This surface abuts against the surface of the origin of gastrocnemius lateralis on the parafibula, as reconstructed in Figs. 43A, B, 44A. It seems that in studied multituberculates soleus was not completely separated from gastrocnemius lateralis. Musculus plantaris possibly was not present in the studied multituberculates, and this condition is also characteristic of monotremes (Haines 1942). On the calcanea from the Late Cretaceous of North America (Fig. 55F–I) and in ?*Eucosmodon* (Fig. 56H–J), which are better preserved than in studied Mongolian multituberculates, there is no tuberosity on the tuber calcanei for its insertion.

Muscles of the foot
Figs. 40, 42, 43

Long extensors. – Musculus tibialis anterior originated on the large part of the laterocaudal surface of the tibia and craniomedial surface of fibula and at the prominent, triangular, hooked processes on the lateral side of the proximal ends of these bones in *Kryptobaatar* and *Nemegtbaatar* (Figs. 2, 17F, H, 40B, 43A–C). In contemporary mammals (e.g., *Antechinus* and in *Meriones*, personal observations) this muscle originates from the proximal one-third of the tibia, from the fascia of peroneus longus, and on the proximal part of the fibula. In *Kryptobaatar* and *Nemegtbaatar*, between the two tendons that originated from the lateral hooked processes of the tibia and fibula, a tendon of ext. digitorum longus passed onto the femur. Its course is more apparent in *Kryptobaatar*

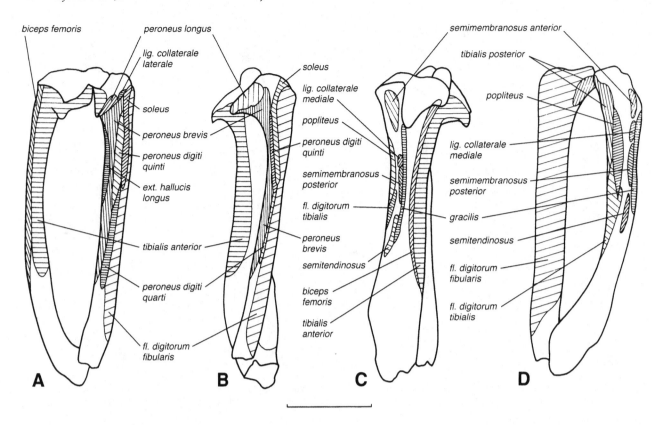

Fig. 43. Left tibia and fibula. Reconstruction of muscle attachments based on surface topography of bones in *Nemegtbaatar gobiensis* (ZPAL MgM-I/110). □A. Lateral. □B. Caudal. □C. Cranial. □D. Medial views. Scale bar 5 mm.

than in *Nemegtbaatar*. In the latter these hooked processes are very close to each other, but area of the origin of the tendon of ext. digitorum longus at the lateral epicondyle of the femur (Fig. 42C) demonstrates the existence of this tendon. In *Kryptobaatar* m. tibialis anterior inserted on a fovea at the mediodorsal surface of the entocuneiform and possibly on the proximal end of Mt I (Fig. 7E). Extensor digitorum longus originated from the lateral epicondyle of the femur (Fig. 42C). In most modern mammals there are two sites of insertion of this muscle. As a rule it inserts directly on the ungual phalanges or on the distal ends of Ph 2 of D II – D IV. In *Kryptobaatar* this muscle possibly inserted directly onto the ungual phalanges.

Extensor hallucis longus is large in contemporary climbing mammals. In non-climbing mammals it is small, and there is no trace of its origin on neither the fibula nor the tibia. We have not found such a trace in *Nemegtbaatar* or *Kryptobaatar*, but this does not exclude the possibility of its presence, and we tentatively reconstruct it in Fig. 43A.

Musculus peroneus longus originated in *Nemegtbaatar* and *Kryptobaatar* from the lateral side of the triangular hooked process of the fibula, immediately lateral to the origin of soleus (Fig. 43A, B). This muscle lies in a fissure found only on the left fibula in *Kryptobaatar* (to the right and below the number 12 in Fig. 2B). In other specimens this part of fibula is broken. Peroneus longus in *Kryptobaatar* appar-

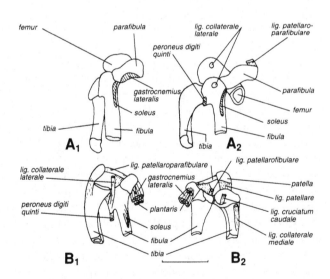

Fig. 44. Reconstruction of origin of m. gastrocnemius and tendons on parafibula in multituberculates and marsupials. □A. *Kryptobaatar dashzevegi*; A₁, A₂ lateral and medial view. □B. *Antechinus stuarti*; B₁, tibia in lateral, fibula in cranial view; B₂, tibia in medial, fibula in caudal view. Scale bar 5 mm.

ently passed through the peroneal groove in the calcaneum onto the plantar side of Mt V (number 11 in Fig. 2B), but we were unable to reconstruct its further course. However, we reconstructed the course of the tendon of m. peroneus longus in ?*Eucosmodon* (see Fig. 57B and 'Functional anatomy'). Musculus peroneus brevis originated from the medial side of the fibula, starting near the origin of peroneus longus and covering about ⅕ of the length of this bone. Peroneus brevis passed through a sulcus situated cranial to peroneus longus. In *Kryptobaatar* it inserted at a bulge at the latero-proximal end of Mt V, and on this basis we reconstruct it in *Nemegtbaatar* (Fig. 40B). Peroneus digiti quarti and peroneus digiti quinti originated in *Nemegtbaatar* and *Kryptobaatar* on the lateral surface of the fibula between soleus and peroneus brevis, peroneus digiti quarti more distally than peroneus digiti quinti (Fig. 43A, B). The sites of insertions of these muscles were not found, and the insertion of peroneus digiti quinti is only tentatively reconstructed in Fig. 40B.

Long flexors. – Musculus tibialis posterior, fl. digitorum tibialis and fl. digitorum fibularis in *Kryptobaatar* and *Nemegtbaatar* originated on the internal surface of the tibia and fibula and from the interosseous membrane (Fig. 43C, D). Flexor digitorum fibularis originated laterally and fl. digitorum tibialis medially. In extant mammals, a part of this latter muscle usually originates distal to popliteus. Only in some contemporary mammals a small slip of fl. digitorum tibialis originates outside m. popliteus. Such a slip probably also existed in *Nemegtbaatar*. In *Kryptobaatar* this part of the tibia is poorly preserved. In contemporary mammals the tendon of fl. digitorum fibularis passes medial to the calcaneum and then divides into the five tendons for five ungual phalanges. Flexor digitorum tibialis merges with the common tendon of the fl. digitorum fibularis or passes towards the internal surface of the calluses of the foot. In *Kryptobaatar* m. tibialis posterior probably inserted on the plantar surface of the cuboid.

We were unable to identify short extensors and short flexors in the multituberculates we studied.

Anatomical comparisons

In the present chapter we do not cite the 19th-century papers referring to multituberculate postcranial fragments (some of which were wrongly allocated). We refer readers to a thorough review of previous descriptions of multituberculate postcranial material by Krause & Jenkins (1983).

Proportions of the body

As appears from our reconstructions of the skeleton and body (Figs. 45, 61), the multituberculates differ in propor-

tions from most small modern therian mammals. Because of limb abduction, their body was held lower to the ground than in mammals with more parasagittally oriented limbs and is similar in proportions to that reconstructed for *Gobiconodon* (Jenkins & Schaff 1988, Fig. 1). A characteristic feature of the multituberculate body is the relatively large head and short neck. For example, in *Kryptobaatar* the length of the skull is 34 mm, which corresponds to 600% of the length between pre- and postzygapophyses of L7. In *Nemegtbaatar* the length of the skull is 40 mm and the width 31 mm, which corresponds to 570% and 440%, respectively, of the length between pre- and postzygapophyses of L7. In marsupials, the same proportions are 563% and 287% in *Didelphis marsupialis* and 360% and 220% in *Phalanger maculatus*; in rodents the proportions are 425% and 300% in *Spermophilopsis leptodacytlus* and 445% and 210% in *Rattus norvegicus*; while in *Gobiconodon ostromi* (measured in the reconstruction in Fig. 1 of Jenkins & Schaff 1988) the length of the skull is 660%.

In comparison to the length of the femur, the length of the skull is 140% in *Kryptobaatar*, 126% in *Didelphis marsupialis*, 97% in *Phalanger maculatus*, 120% in *Spermophilopsis leptodacytlus*, 125% in *Rattus norvegicus* and 181% in *Gobiconodon*.

It follows that the multituberculate skull is relatively longer than in marsupials and rodents measured by us but shorter than in *Gobiconodon* (if the latter was correctly reconstructed by Jenkins & Schaff 1988); it is, however, notably wider than in measured extant mammals (unknown in *Gobiconodon*).

Hyoid apparatus

We have found an incomplete hyoid apparatus (previously unknown in multituberculates) in *Chulsanbaatar*. Because of its incompleteness (only apparent stylohyoid bones have been preserved) it is difficult to compare it with those of other mammals (Gasc 1967).

Vertebral column

Cervical vertebrae

The atlas, previously unknown in multituberculates has been found in *Chulsanbaatar* and *Nemegtbaatar*. The transverse foramen is absent in Monotremata and most Marsupialia but occurs in extant Eutheria. Among Late Cretaceous eutherians, it is absent in *Asioryctes* but present, although very small, in *Barunlestes* (Kielan-Jaworowska 1977, 1978). The transverse processes are imperforate in *Chulsanbaatar*. In *Nemegtbaatar*, there is a foramen similar in position to that of the transverse foramen in *Barunlestes*, but smaller, possibly too small to transmit the arteria vertebralis, and recog-

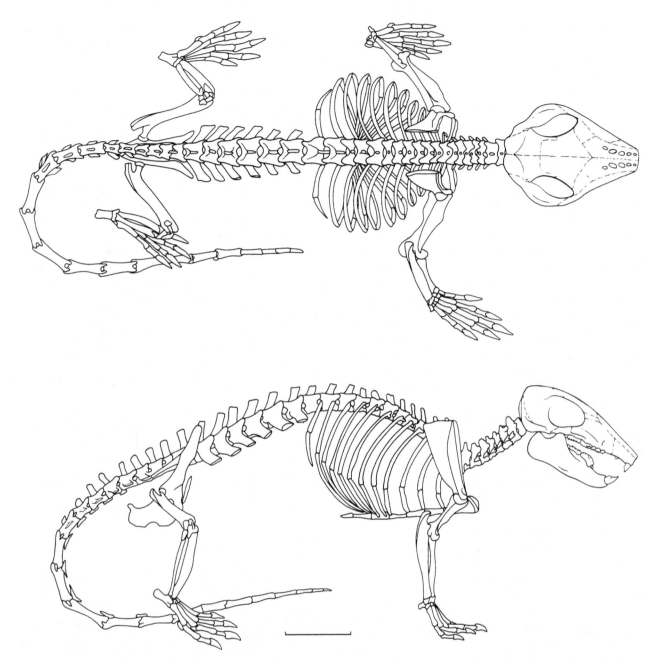

Fig. 45. Reconstruction of skeleton of *Nemegtbaatar gobiensis*, based in part on *Kryptobaatar dashzevegi* and *Chulsanbaatar vulgaris*, in dorsal and lateral views. The manus, carpus, shoulder girdle (except for the ventral part of the scapulocoracoid), sternum, most of the ribs, part of the thoracic vertebrae and part of the caudal vertebrae have not been preserved and are reconstructed. Scale bar 20 mm.

nized by us as a ?vascular foramen. In *Barunlestes*, there is a distinct sulcus arteriae vertebralis extending craniodorsally from the anterior transverse foramen (Kielan-Jaworowska 1978, Pl. 2: 1a), which is not obvious in *Nemegtbaatar*. The transverse foramen in the multituberculate atlas was apparently absent.

The incomplete axis has been found for the first time in multituberculates in *Nemegtbaatar* and *Chulsanbaatar*. Judging from the preserved parts, the spinous process (most complete in the juvenile *Nemegtbaatar*, Fig. 9A, C) was

relatively small, in particular in relation to a large and heavy head. In multituberculates the cervical vertebrae are short with relatively high arches.

There are seven cervical vertebrae in extant mammals with few exceptions (Lessertisseur & Saban 1967a, Table I). The number of cervical vertebrae is not known for triconodonts and other Mesozoic mammals (Jenkins & Parrington 1976, Lillegraven *et al.* 1979). In the galeosaurid cynodont *Thrinaxodon*, Jenkins (1971a, p. 49) found that 'The determination of the number [of cervical vertebrae] is complicated by the fact that *Thrinaxodon*, unlike most mammals, retained cervical ribs, the absence of which is a primary feature of mammalian cervical vertebrae'. Jenkins concluded, however, that *Thrinaxodon* had seven cervical vertebrae. In the Liassic tritylodontid cynodont *Oligokyphus* (Kühne 1956) there are also seven cervical vertebrae.

The complete cervical series associated with the first thoracic vertebrae is preserved in *Nemegtbaatar* (ZPAL MgM-I/82) and (without thoracic vertebrae) in an unidentified taeniolabidoid (ZPAL MgM-I/165, Kielan-Jaworowska 1989). In addition, two specimens of *Chulsanbaatar* preserve the damaged cervicals. In most mammals the seventh vertebra differs from the preceding ones in lacking the transverse foramen (e.g., Lessertisseur & Saban 1967a; Getty 1975; Evans & Christensen 1979). In *Nemegtbaatar*, although the complete transverse processes on C7 have not been preserved, the groove between the broken ventral and dorsal branches of the process indicates the presence of a transverse foramen. In an unidentified taeniolabidoid (Kielan-Jaworowska 1989, Text-fig. 1, Pls. 15, 16:1) the transverse process on C7 is long and perforated by a transverse canal, rather than reduced to a single-pronged process. The caudal costal fovea that occurs on C7 in many mammals is not obvious in *Nemegtbaatar* but is preserved in the unidentified taeniolabidoid noted above. This and the broken-off heads of the first ribs cemented to the cranial costal foveae of T1 in *Nemegtbaatar* (Fig. 9A–C) demonstrate that there were seven cervical vertebrae in multituberculates, as in most mammals. Jenkins & Parrington (1976) do not mention the transverse foramina on the cervical vertebrae in a triconodont *Megazostrodon*, although the foramen appears to be present (Fig. 6b of their paper), at least on C5.

The transverse foramen on C7, characteristic of multituberculates, occurs in *Tachyglossus*, where it is smaller than on the sixth vertebra. It is rarely present in extant eutherian mammals. In *Lepus*, for example, the transverse process, situated opposite the anteriormost part of the body, is pierced by a transverse foramen (personal observations); such a foramen occurs also in *Homo* (Davies & Davies 1962). The caudal costal fovea on C7 is, in *Lepus*, very small, and the first rib articulates mostly with the cranial fovea of T1, as is characteristic also of *Nemegtbaatar*. However, the shape of the body of C7 differs in multituberculates from that in *Lepus* and most extant small eutherian mammals. In *Lepus* the body of C7 differs from C6 in having a trapezoidal shape (ventral view), while C6 is rectangular with prominent inferior (ventral) lamellae (Howell 1926). In multituberculates the difference between the shape of C6 and C7 is less dramatic, as the inferior lamella (for the origin of m. longus capitis and longus atlantis and insertion of m. longus colli) is not developed on C6 and both vertebrae have a roughly trapezoidal shape in *Nemegtbaatar*; they are more rectangular in the unidentified taeniolabidoid (Kielan-Jaworowska 1989).

Cervical ribs

In cynodonts the cervical ribs are slender, directed laterocaudally and ventrally, and do not overlap each other (Kühne 1956; Jenkins 1971a; Kemp 1980). Fragmentary cervical ribs, not in articulation with the vertebrae, have been found in the Triassic morganucodontid triconodont *Megazostrodon*; on this basis Jenkins & Parrington (1976, Fig. 18) reconstructed slender and ventrally directed cervical ribs in a Triassic triconodont. In modern mammals cervical ribs occur in monotremes (on C2–C7 in *Ornithorhynchus* and *Tachyglossus* and on the axis in *Zaglossus*) and on the axis in one marsupial (*Perameles*), while free ribs may occur rarely on C7 or even more anterior cervicals in a few eutherians (Lessertisseur & Saban 1967a). Cervical ribs are wide, rounded at the ends, and overlap each other in *Ornithorhynchus*; they are more pointed in *Tachyglossus*.

We have found damaged cervical ribs on C2 and C4 in *Nemegtbaatar* and on C2 (uncertain), C5 and possibly also C7 in *Chulsanbaatar* (Figs. 9, 20B). In these taxa they are relatively large and roughly oval in lateral view. Multituberculate cervical ribs are reminiscent of those of monotremes, rather than those of cynodonts and triconodonts. In a taeniolabidoid fam., gen. et sp. indet. (Kielan-Jaworowska 1989, Pls. 15, 16:1), which possibly belonged to an old individual, the transverse processes are very long and pointed, and the ribs are not discernible. It is possible that the cervical ribs were present in young individuals of this taxon and became fused to the transverse processes in old individuals, as we observed in *Tachyglossus*. We conclude that cervical ribs were characteristic of at least some multituberculate taxa.

Thoracic vertebrae, ribs and sternum

The number of thoracic vertebrae and ribs is not known in multituberculates. On the basis of the damaged vertebral column in *Chulsanbaatar* (Fig. 18) we estimate that there were 13 thoracic vertebrae in this genus. Several broken ribs preserved in *Nemegtbaatar* and *Chulsanbaatar* do not allow an estimation of their number.

The manubrium of the sternum and one sternebra have been preserved only in an unidentified taeniolabidoid ZPAL MgM-I/165 (Kielan-Jaworowska 1989). This specimens shows that multituberculates had ossified sternebrae.

Lumbar vertebrae

The number of lumbar vertebrae in multituberculates is uncertain. Krause & Jenkins (1983) stated that in *Ptilodus* at least seven, but possibly eight, lumbar vertebrae are present. We tentatively conclude that there were seven lumbar vertebrae in multituberculates, on the basis of a poorly preserved specimen of *Chulsanbaatar* (Fig. 18), although the diaphragmatical vertebra cannot be recognized with any certainty.

The multituberculate lumbars differ from those of modern therians, monotremes and some cynodonts (but not of the tritylodontids) in lacking the anapophyses (Jenkins 1970b; Sues 1985). They also differ from those in all extant mammals in having poorly developed metapophyses that are not well separated from the prezygapophyses. We have found very long transverse and spinous processes of the lumbar vertebrae in Asian multituberculates (Table 2); it is not known whether similarly long processes were characteristic of all multituberculates.

Sacrum

The number of sacral vertebrae in mammals varies considerably. There are two sacrals in monotremes and marsupials (but only one in *Perameles*), most commonly three in eutherian mammals, but for example four to five in domestic mammals, nine in most recent xenarthrans and as many as 17 in glyptodonts (Lessertisseur & Saban 1967a; Getty 1975; Starck 1979).

Granger & Simpson (1929) mentioned a crushed fragment of a sacrum of the taeniolabidoid ?*Eucosmodon*, consisting of at least two fused vertebrae. Krause & Jenkins (1983) stated that there are four sacral vertebrae in the ptilodontoid *Ptilodus kummae*. In the material studied by us, incomplete sacra are preserved in seven specimens belonging to *Kryptobaatar*, *Chulsanbaatar*, *Nemegtbaatar*, *Sloanbaatar* and *Kamptobaatar* (the latter poorly preserved and not described here). We support the conclusion of Krause & Jenkins (1983) that the multituberculate sacrum consisted of four fused vertebrae. In all of the studied genera the sacrum is long, and the second sacral vertebra is longer than the first one.

The first two sacral vertebrae articulate with the ilium in *Kryptobaatar* and in *Chulsanbaatar*, but this is not known in *Nemegtbaatar*. Because of the elongation of S2, the iliosacral contact in *Kryptobaatar* is relatively very long. This contact is, at least in *Kryptobaatar*, more dorsoventral than mediolateral, and in the latter genus the sacrum appears to be more firmly synostosed with the ilia than in *Chulsanbaatar* and *Nemegtbaatar*.

Caudal vertebrae

We have found three anterior caudal vertebrae in *Kryptobaatar* and two isolated caudals in *Catopsbaatar*. Caudal vertebrae in *Nemegtbaatar* have not been preserved, but on the basis of distinct scars for the origin of m. sacrocaudalis dorsalis (Fig. 36B) on the last four lumbar vertebrae (damaged on anterior lumbars), we conclude that the tail was possibly longer than the rest of the body in *Nemegtbaatar*.

Pectoral girdle and forelimb

Scapulocoracoid, interclavicle and clavicle

All described multituberculate scapulocoracoids (Simpson 1928; McKenna 1961; Clemens & Kielan-Jaworowska 1979; Jenkins & Weijs 1979; Krause & Jenkins 1983) are incomplete. The multituberculate scapulocoracoid has been, until recently, considered to be narrow, have a trough-like infraspinous fossa and a laterally reflected cranial border, and lack a supraspinous fossa. It appears from the fragmentary scapulocoracoids preserved in *Kryptobaatar* and *Nemegtbaatar* that the infraspinous fossa widens dorsally rather than being trough-like. Clemens & Kielan-Jaworowska (1979, p. 117) pointed out that the supraspinous fossa 'is absent during part of the embryological development of therians. For example, the foetus of *Didelphis* possesses only an infraspinous fossa until it reaches 7.5 mm in crown–rump length (Cheng, 1955)'. Jenkins & Weijs (1979, p. 407) stated with respect to the multituberculate scapulocoracoid: 'The laterally reflected anterior border lies above the middle of the glenoid, and thus the musculature arising from the anterior aspect of the border (presumably homologous with the supraspinatus) passed directly ventrad to the greater tuberosity.' We recognize, however, in *Kryptobaatar* and *Nemegtbaatar* the narrow fossa lying to the rear of the cranial (anterior) border as an incipient supraspinous fossa (Figs. 5, 12, 13I, 28C). Such an incipient supraspinous fossa is also apparently present in front of the spine in an unidentified multituberculate scapulocoracoid from the Lance Formation, figured by McKenna (1961, Fig. 3). The scapulocoracoid of *Nemegtbaatar* shows that the cranial border is developed ventrally as a prominent keel and is more reflected laterally than previously known.

The supraspinous fossa does not occur in monotremes (Gregory & Camp 1918), nor in Liassic triconodonts (Jenkins & Parrington 1976). It has been found, however, by Jenkins & Schaff (1988) in a scapula attributed to the Early Cretaceous triconodont *Gobiconodon*, where it is very large (similar to that of therians). Gregory & Camp (1918) described an incipient supraspinous fossa in *Cynognathus* (see Jenkins 1971a, for a review of literature on the shoulder girdle in cynodonts). Kühne (1956) does not mention its presence in the advanced tritylodontid *Oligokyphus*, in which the scapular part of the scapulocoracoid is narrow, as in multituberculates, while the coracoid part is very extensive. It seems to us that the narrow fossa reconstructed by Kühne (1956, Fig. 52B) on the right side of the cranial view of the scapulocoracoid, might correspond to an incipient supraspinous fossa.

The multituberculate coracoid is developed as a small process on the ventral angle of the scapula and is not bent medially as in therian mammals. The ventral surface of the coracoid prolongs the longitudinal diameter of the glenoid fossa and thus reduced the flexor–extensor mobility of the humeral joint. We should make it clear that in animals with abducted limbs, flexion–extension of the forelimb does not correspond to protraction–retraction as in mammals with parasagittal limbs (referred to further as 'parasagittal'; see 'Functional anatomy') but is more complicated, as at the same time abduction–adduction takes place. In multituberculates there is no supraglenoid tubercle characteristic of the Theria (Fig. 28). The multituberculate acromion described herein is peg-like and 'embraces' the humeral head craniolaterally. One can visualize a clavicle abutting against it as reconstructed by Jenkins & Weijs (1979, Fig. 12c) but situated lower than in their reconstruction, in a position rather similar to that in *Didelphis* (Jenkins & Weijs 1979, Fig. 12d). We reconstruct the multituberculate scapulocoracoid as leaning backwards as in therian mammals (Fig. 45).

Until recently the multituberculate interclavicle and clavicle were not known with certainty (see Krause & Jenkins 1983, for review). Sereno & McKenna (1990) reported the presence of these elements in a multituberculate from the Late Cretaceous of Mongolia, but the material has yet to be described. We have not found interclavicles and clavicles in the material studied.

Humerus

Numerous incomplete multituberculate humeri have been described or figured by Gidley (1909), Simpson (1928a), Deischl (1964), Sahni (1972), Jenkins (1973), Kielan-Jaworowska & Dashzeveg (1978), Krause & Jenkins (1983), Kielan-Jaworowska (1989) and Bleefeld (1992); two complete multituberculate humeri (both broken and glued together) were described by Kielan-Jaworowska & Qi (1990). The multituberculate humerus is characterized by a spherical head, the lesser tubercle only slightly smaller than the greater one, wide intertubercular groove, teres tuberosity crescent-shaped; posterior crest, extending in dorsal aspect from the head to ectepicondylar flange; a robust and wide distal end, with very convex radial and ulnar condyles, entepicondylar foramen and a strong degree of twisting. The multituberculate humerus is reminiscent of that in primitive triconodonts in having almost subequal greater and lesser tubercles, a crescent-shaped teres tuberosity and wide distal end (Jenkins & Parrington 1976). These characters occur also in cynodonts (Jenkins 1971a) and monotremes (personal observations); in both these groups, however, the lesser and greater tubercles are less clearly defined, and the humeri are more robust than in multituberculates.

Kielan-Jaworowska & Dashzeveg (1978) possibly overestimated the length of the humerus in *Tugrigbaatar* and underestimated the width of the distal part. The humeri of ?*Lambdopsalis* (Kielan-Jaworowska & Qi 1990) and those of the North American multituberculates belonging to Ptilodontoidea and Taeniolabidoidea (Krause & Jenkins 1983) have an expanded distal end. If this holds for all multituber-

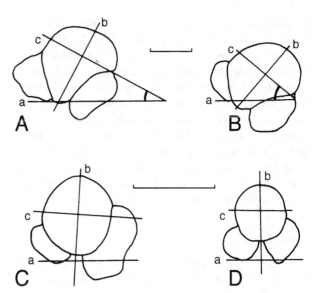

Fig. 46. Diagram illustrating proximal views of left humeri and degree of twisting in multituberculates (A, B), eutherians (C) and marsupials (D). The greater tubercle is to the right. □A–B. ?*Lambdopsalis bulla*, IVPP V9051 and IVPP V8408, respectively. □C. *Rattus norvegicus* ZIN 83. □D. *Marmosa* sp., ZIN 1110. To establish the angle of twisting, the humerus was placed vertically, and its distal axis, across the epicondyles (a), corresponding to the axis of the flexor–extensors in the ulnar joint, was drawn with the camera lucida. Then on the same drawing the humeral head and the line across the proximal end of the humerus, between the middle of the intertubercular groove and the middle of the caudal margin of the humeral head (b), were drawn. The perpendicular to the latter line (c) corresponds to the flexor–extensor excursion in the shoulder joint; the angle of twisting lies between it and the axis of the flexor–extensors of the distal end. In C and D the lines a and c are subparallel, and there is almost no twisting. Scale bars 5 mm.

culates, the reconstruction of the *Tugrigbaatar* humerus should be corrected.

Two right humeri of ?*Lambdopsalis bulla* allow us to establish the degree of relative twisting of the proximal and distal ends for this genus (Fig. 46; see also Simpson 1928b, Fig. 52). In small therian mammals measured by us the angle of twisting is about 5°, in *Tachyglossus aculeatus* 45° and in *Eozostrodon* about 50° (Jenkins & Parrington 1976). In ?*Lambdopsalis* IVPP V8408 it is 38° while in IVPP V9051 it is 24°; this difference is apparently due to inaccuracy in gluing the parts of both humeri. It seems to us that IVPP V8408 is closer to the natural condition and that the angle was about 33–35°. The intertubercular groove is very wide in multituberculates; it is 32% of the width of the proximal epiphysis in *Nemegtbaatar* and 28–31% in ?*Lambdopsalis*, while in small modern rodents it is only 14–23%.

Radius and ulna

Proximal parts of multituberculate ulnae have been described by Jenkins (1973) and incomplete radii and ulnae by Jenkins & Krause (1983). The proximal part of the left radius, preserved together with an almost complete ulna (with a broken olecranon) in *Nemegtbaatar* (ZPAL MgM-I/81), is the most complete multituberculate forearm known so far (Fig. 14A). The distal end of our ulna, although damaged, resembles that in small therian mammals and strongly differs from that in *Tachyglossus.* Therefore we suggest that the rotation between the distal parts of radius and ulna in multituberculates was around the styloid process, as in primitive therian mammals (Lewis 1989). Krause & Jenkins (1983) described the multituberculate radius as mediolaterally compressed. The proximal and distal parts of the radii preserved in *Nemegtbaatar* are craniocaudally compressed.

Manus

An incomplete manus of *Ptilodus* was described by Krause & Jenkins (1983). In this paper we describe four tentatively identified carpal bones of *Nemegtbaatar.* These, together with the material described by Krause & Jenkins, show that there were three bones in a proximal row in the multituberculate carpus and five (including the praepollex) in a distal row, but the centrale has not been found as yet. As the manus has not been preserved in multituberculates studied by us, its reconstruction in Fig. 45 is entirely tentative.

Pelvic girdle and hind limb

Pelvis

An incomplete pelvis of *Ptilodus* was described by Gidley (1909), who suggested that multituberculates probably had marsupial (epipubic) bones. Kielan-Jaworowska (1969) found multituberculate epipubic bones in *Kryptobaatar* (see also Fig. 2 in this paper).

Granger & Simpson (1929) described an incomplete pelvis of ?*Eucosmodon* and stated that it differs from those of therian mammals in having the acetabulum open dorsally and the ischia meeting each other at an acute angle. An open acetabulum is characteristic also of triconodonts (Jenkins & Parrington 1976), the tritylodontid *Oligokyphus* (Kühne 1956) and some arboreal (gliding) marsupials (Elftman 1929), while in monotremes it is only partly covered dorsally (see 'Functional anatomy' for interpretation).

Deischl (1964) described several incomplete multituberculate pelves. Kielan-Jaworowska (1969, 1979) demonstrated that in *Kryptobaatar* the ischial arc is very small, limited only to the dorsal part of the ischia, the ischial tuber is developed as a parabolic process, strongly protruding dorsally, and parts of the pubes and ischia are firmly fused ventrally to form a keel. Krause & Jenkins (1983) found the keel of the ischiopubic symphysis in North American genera belonging to both the Taeniolabidoidea and Ptilodontoidea. They also noted the presence of a postobturator foramen within the ischiopubic symphysis in *Ptilodus*, ?*Eucosmodon* and ?*Mesodma* that does not open to the pelvic cavity. In the four taeniolabidoid genera described or discussed here (*Kryptobaatar, Chulsanbaatar, Sloanbaatar* and *Kamptobaatar*) the pelvis is narrow with a keel-like ischiopubic symphysis, and a postobturator notch (or foramen) was possibly present within the symphysis, at least in *Kryptobaatar* (Fig. 2). The only pubis of *Nemegtbaatar* (Fig. 16F) has been distorted, and the ventral keel has not been preserved. As, however, the keel occurs in all the genera belonging to both the Ptilodontoidea and Taeniolabidoidea in which this region is known, we assume that it occurred also in *Nemegtbaatar* and was characteristic of Cimolodonta and possibly is a multituberculate autapomorphy.

The multituberculate pelvis differs from those of therian mammals in being deeper. This is related to the abducted position of the multituberculate hind limbs and the origin of femoral adductors ventral to the acetabulum, rather than caudal, as in therian mammals (see 'Functional anatomy' and Figs. 48, 49). A deep pelvis is also characteristic of monotremes (Gregory & Camp 1918; Jouffroy & Lessertisseur 1971), but the multituberculate pelvis differs greatly from that of monotremes in having a very narrow ischial arc; whereas in monotremes, in relation to oviparity, the ischial arc is widely open, U-shaped.

In multituberculates the iliosacral angle is very large in comparison to that in extant therian mammals (Fig. 37B_2, D_1 and Table 1). In extant mammals with 'parasagittal' hind limbs (see 'Functional anatomy') the iliosacral angle varies with the body mass of the animal (Lessertisseur 1967). In

Table 1. Iliosacral angle in multituberculates and extant mammals.

Cat. No.	Species	Ang	Hab
81	*Nemegtbaatar gobiensis*	33°	t
85	*Chulsanbaatar vulgaris*	37°	t
41	*Kryptobaatar dashzevegi*	36°	t
✻	*Ornithorhynchus anatinus*	40°	aq
31024	*Tachyglossus aculeatus*	38°	f, t
13951	*Antechinus stuarti*	17°	t
1110	*Marmosa* sp.	11°	s
192	*Neomys fodiens*	9°	aq, t
1380	*Elephantulus rozeti*	16°	r
800	*Sciurus persicus*	19°	s
175	*Citellus fulvus*	18°	t
206	*Citellus xanthoprymnus*	9°	t
59	*Spermophylopsis leptodactylus*	14°	t
215	*Dryomys nitedula*	15°	s
221	*Glis glis*	14°	s
83	*Rattus norvegicus*	16°	t
314	*Mesocricetus brandti*	10°	t
311	*Cricetulus migratorius*	9°	t
428	*Meriones tamariscinus*	13°	t
117	*Allactaga jaculus*	19°	r

Cat. No. (catalogue numbers) refer to ZPAL MgM-I/ collection for the first three taxa; the remainder to ZIN, Laboratory of Mammals collection. Ang = angle; Hab = habits; aq = aquatic; f = fossorial; r = ricochetal; t = terrestrial; s = scansorial; ✻ = after Lessertisseur 1967.

small mammals this angle ranges from 9° to 19°. In multitu-berculates this angle is 35–37° (measured by the method of Huxley 1879, rather than Le Damany 1906, because of the lack of complete skeletons; see also Lessertisseur 1967 and Fig. 37B$_2$ and D$_1$ herein). This angle in multituberculates is of roughly the same value as in *Tachyglossus* (38°), while in *Ornithorhynchus* it is 40°. A large iliosacral angle in mono-tremes and multituberculates is possibly a plesiomorphic feature.

Jenkins & Schaff (1988) reconstructed the pelvis of the Cretaceous triconodont *Gobiconodon* (in which the ventral part has not been preserved) as very shallow, similar to that of therians. As the hind limbs in *Gobiconodon* were abducted, we speculate that the triconodont pelvis was possibly deeper, similar to that of multituberculates, and the iliosacral angle larger.

Femur

The femur is the stoutest bone in the multituberculate skel-eton, and several fragmentary and complete femora belong-ing to different genera are known (Gidley 1909; Deischl 1964; Simpson & Elftman 1928; Granger & Simpson 1929; Sloan & Van Valen 1965; Krause & Jenkins 1983; and this paper). The multituberculate femur is characterized by a head with an extensive articular surface and a long neck forming an angle of 50–60° to the shaft; a prominent greater trochanter ex-tending beyond the head; a prominent plate-like lesser tro-chanter, convex lateroproximally and concave medio-distally, strongly protruding ventrally and placed at the point of confluence of the greater trochanter with the neck; ab-sence of the third trochanter; in dorsal aspect a distinct subtrochanteric tubercle (see 'Terminology'); small distal condyles; and a shallow trochlea.

The trochanteric (digital) fossa has been described by Simpson & Elftman (1928) and Krause & Jenkins (1983) as divided into a proximal part situated between the two tro-chanters and a distal part, fissure-like in shape, placed lateral to the lesser trochanter. Granger & Simpson (1929, p. 641) stated: 'Lateral to the lesser trochanter and between it and the gluteal crest is another equally definite and larger [than the digital fossa] elongated fossa which does not have any certain separate homologue on any other femur compared by us.' We believe that the latter fossa is not related to the trochan-teric fossa and we designate it (see 'Terminology') the post-trochanteric fossa.

The multituberculate femur is very different from that of monotremes, triconodonts and docodonts (Vialleton 1924; Parrington 1961; Jenkins & Parrington 1976; Jenkins & Schaff 1988; Henkel & Krusat 1980; Krusat 1991), in which the femoral neck is short and wide, the greater trochanter does not extend beyond the head, and both trochanters are triangular. These trochanters are very different in shape and position from those in multituberculates and arise sym-metrically (in ventral and dorsal views) on both sides of the

head. The morphology of the femora of therian mammals is diverse, but in general the head is placed on a well-developed neck and projects craniomedially as in multituberculates. The greater trochanter, which is very prominent in multitu-berculates, is variably developed in therians, usually lower than, or as high as the head (Greene 1935; Lessertisseur & Saban 1967b; Getty 1975; Starck 1979; and personal observa-tions). The lesser trochanter in multituberculates as charac-terized above is very different from that in therians. The subtrochanteric tubercle characteristic of multituberculates does not occur in other mammals, to our knowledge. The function of this tubercle is not clear to us. Simpson & Elft-man (1928) regarded it as the insertion of m. iliocapsularis (which they regarded as a synonym of iliofemoralis). How-ever, iliocapsularis is a synonym of capsularis, and iliofemo-ralis of reptiles is homologous to gluteus (Romer & Parsons 1986). As m. capsularis in modern mammals inserts more medially, close to the lesser trochanter, it seems more prob-able that the discussed tubercle corresponds to the insertion of lig. iliofemoralis.

Tibia, fibula and parafibula

Fragmentary or complete tibiae, belonging to several ptilo-dontoid and taeniolabidoid genera, are known (Gidley 1909; Granger & Simpson 1929; Deischl 1964; Krause & Jenkins 1983). For a long time the fibulae were incompletely known, and the first complete one (belonging to *Ptilodus*) was de-scribed by Krause & Jenkins (1983). We described above tibiae and fibulae preserved together in *Kryptobaatar*, *Nem-egtbaatar* and *Chulsanbaatar*.

In mammals with 'parasagittal' (see 'Functional anato-my') limbs, the craniocaudal diameter of the tibia is much greater than the mediolateral (Kummer 1959a, 1959b; Soko-lov *et al.* 1974). In multituberculates, in contrast, the medio-lateral diameter is relatively large in respect to the cranio-caudal, which is possibly related to the fact that because of the abducted position of their limbs, the stress on the tibia during the propulsive phase is directed medially. In *Krypto-baatar* (ZPAL MgM-I/41) the craniocaudal diameter of the tibia is 54% of the mediolateral diameter, in *Nemegtbaatar* (ZPAL MgM-I/110) 60% and in ?*Eucosmodon* 63% in the proximal part and 65% in the distal part. In modern therian mammals these values are 81% in *Marmosa* sp. (ZIN 1110), 107% in *Elephantulus rozeti* (ZIN 1380) and 105–150% in different species of rodents cited in Table 2.

The multituberculate tibia is characterized by a markedly asymmetrical proximal end, with a small medial and a large lateral facet. The prominent, hook-like triangular process overhangs the shaft laterally and bears a facet for the fibula. Deischl (1964) noted that the distal end of multituberculate tibia is similar to that of *Didelphis*, where there is a very prominent medial malleolus and a large, relatively flat lateral condyle. We observed similar structure in other marsupials, including *Antechinus stuarti* (Dasyuridae), *Trichosurus vul-*

pecula and *Phalanger maculeatus* (Phalangeridae), as well as in *Kryptobaatar, Nemegtbaatar* and an unidentified multituberculate from the Hell Creek Formation (Fig. 55A). In ?*Eucosmodon* the distal end of the tibia differs from that of Asian multituberculates and from a tibia from the Hell Creek Formation in having a small, flat medial condyle developed in addition to the medial malleolus (Fig. 56F). Krause & Jenkins (1983, Fig. 22) referred to the medial malleolus in the tibia of ?*Eucosmodon* as the medial tibial condyle. This notation may be justified as, in fact, on the cranial and caudal sides of the medial malleolus there are articular surfaces for the astragalus (see 'Functional anatomy').

The fibular head bears on its lateral part a hook-like triangular process for articulation with the above-mentioned hook-like process on the tibia. In lateral view these processes are aligned. The fibular shaft appears relatively more robust (with respect to that of the tibia) in *Nemegtbaatar* and *Chulsanbaatar*, than in *Kryptobaatar*. Although the fibula in *Ptilodus* has been preserved in an articulated skeleton, Krause & Jenkins (1983) were not able to establish its orientation with certainty. The specimens described here show that the fibula in multituberculates does not participate in the knee joint and is situated caudal to the tibia in the proximal part, but lateral in the distal part.

In the Triassic triconodont *Erythrotherium* the fibula participates in the knee joint and is placed lateral to the tibia, arranged obliquely craniodorsally to ventrocaudally (Jenkins & Parrington 1976). In the Early Cretaceous triconodont *Gobiconodon* the exact position of the fibula in relation to the tibia is not known, but Jenkins & Schaff (1988, Fig. 1) reconstructed it subparallel and lateral (slightly laterocaudal) to the tibia. In monotremes the large fibula is very different from that in multituberculates, and it is placed entirely lateral to the tibia (Haines 1942). The respective sizes of tibia and fibula in extant therian mammals vary (Lessertisseur & Saban 1967b). As demonstrated by Barnett & Napier (1953b), in marsupials (but not in the Peramelidae) there is generally some degree of femorofibular contact, which is absent in Eutheria (Barnett & Napier 1953a). In both groups, however, irrespective of the differences in the size of the fibula, this bone is placed more laterally with respect to the tibia than in multituberculates.

Krause & Jenkins (1983) described the first multituberculate parafibulae, found in association (but not in articulation) with the hind limbs of *Ptilodus*. We have found the parafibulae in articulation, although slightly displaced, in *Kryptobaatar* and *Chulsanbaatar* and an apparent fragment of the parafibula in *Nemegtbaatar* (Figs. 2B, 3A, B, 17F, H, I, 24B, C). It seems probable that the presence of this ossicle was characteristic of multituberculates as a whole. As argued under 'Myological reconstructions' (Fig. 44A), m. gastrocnemius lateralis possibly originated in multituberculates from the parafibula (as in some marsupials), rather than from the femur as in eutherians.

Pes

Isolated bones of multituberculate pedes were described or figured by Gidley (1909), Simpson (1928a), Granger & Simpson (1929), Deischl (1964), Sahni (1972), Bleefeld (1992) and Szalay (1993). Granger & Simpson (1929) reconstructed the multituberculate pes (of ?*Eucosmodon* sp.) with the longitudinal axis extending along the third ray of the foot (we follow Leonardi 1987 in defining the position of the pes after the position of the third ray of the foot – Mt III and D III). They also demonstrated that the cuboid facet in the multituberculate calcaneum is arranged obliquely mediodistally, rather than distally as in all other mammals, and reconstructed the calcaneum not supported distally by any bone. Krause & Jenkins (1983) and Jenkins & Krause (1983) described a nearly complete pes of *Ptilodus* and discussed its function. They argued that the multituberculate femora were not parasagittal, with which we agree. Our reconstructions (Figs. 54 and 57) differ from those of Granger & Simpson (1929, Fig. 29), Krause & Jenkins (1983, Fig. 24) and Szalay (1993, Fig. 9.10) in that we place Mt III at an angle of 30° in respect to the longitudinal axis of the tuber calcanei. In our reconstruction Mt V articulates with the distal margin of the calcaneum medial to the peroneal groove (see 'Functional anatomy').

Multituberculate calcaneum differs from that of the Liassic triconodonts (Jenkins & Parrington 1976) in having a well-developed tuber calcanei, compressed laterally and similar to that in therians. However, Jenkins & Schaff (1988, Fig.1) reconstructed the calcaneum of the Early Cretaceous triconodont *Gobiconodon* with a tuber calcanei similar to that in therians.

At least two types of multituberculate calcanea may be recognized. In one type, represented by Asian calcanea described here, an unidentified multituberculate calcaneum from the Hell Creek Formation (Fig. 55E–G) and an unidentified Late Cretaceous ptilodontoid calcaneum figured by Szalay (1993, Fig. 9.8), the peroneal groove is wide and the peroneal tubercle is directed laterodistally. In the other type, represented by an unidentified multituberculate from the Lance Formation (Fig. 55H, I) and ?*Eucosmodon* sp. (Fig. 56H–J), the peroneal groove is narrow and the peroneal tubercle directed more distally. However, in both types the peroneal groove is very deep and the peroneal tubercle is clearly set off from the calcaneal body; in this respect the multituberculate calcaneum differs from those of other mammals.

If our reconstruction of the multituberculate pes is correct, multituberculates would differ from all mammals in having Mt V articulating with the calcaneum. Among extant reptiles Mt V articulates with the calcaneum (fused to the astragalus) only in *Sphenodon* (Romer & Parsons 1986). Calcaneo–Mt V contact occurs in various extinct tetrapods, e.g., in a few therapsids (gorgonopsians, dinocephalians and dicynodonts); it has also been found in a primitive archosauromorph, some lizards, an early rhynchosaur and some thecodonts (Romer 1956; Bonaparte 1971; Cruickshank 1972; Carroll 1976, 1988; Chatterjee 1978).

In pelycosaurs the fifth tarsal is situated between the calcaneum and Mt V. In bauriamorphs, cynodonts and most therapsids (see, e.g., Schaeffer 1941a, 1941b; Jenkins 1971a; Szalay 1993), there is a gap between the calcaneum and Mt V, which, as demonstrated by Schaeffer (1941a) on the roentgenogram of *Bauria* pes, is not an artefact. Schaeffer argued that this gap may be filled either with a lateral fibrocartilaginous extension from the cuboid, or by a persistent, cartilaginous fifth tarsal. Schaeffer (1941b, p. 5) stated: 'Granger & Simpson (1929) have recorded the presence of a large gap in the same position in the multituberculate *Eucosmodon*, although in this case the fifth metatarsal does not articulate with the extreme lateral border of the cuboid. This gap certainly was not filled by a cartilaginous fifth tarsale.' As argued above, we believe that this gap does not exist in multituberculates.

In the Middle Triassic Manda cynodont examined by Jenkins (1971a) and Szalay (1993), there is a large gap between the calcaneum and Mt V, and the calcaneo-cuboid facet is distal rather than mediodistal. One can visualize that during the evolution leading from therapsids to multituberculates, when the fifth tarsal disappeared, the calcaneum came into contact with Mt V. At the same time the cuboid facet on the calcaneum shifted from nearly distal to mediodistal position. The calcaneo–Mt V contact and mediodistal cuboid facet on the calcaneum are multituberculate autapomorphies. In the evolution leading to therian mammals, when the fifth tarsal disappeared, the cuboid extended laterally, supporting the distal end of the calcaneum, and the cuboid facet acquired a distal position.

In the Mongolian taxa studied by us, the plantar side of the astragalus is poorly exposed, and we have not described the astragalar canal. Szalay (1993, p. 121) stated: 'The unique and arched buttress around the astragalar canal (ac) of multis requires an eventual explanation – it is highly diagnostic.' Granger & Simpson (1929), referring to the process on the astragalus for articulation with the navicular, stated (p. 644): 'The latter process is, of course, homologous with the head of normal primitive therian astragalus, but here it has no neck and is not cut off from the main body of the bone.' This process has also been referred to by Krause & Jenkins (1983) as the astragalar head. We do not follow this usage for the following reasons. Szalay (1993, p. 121) pointed out that: 'The articulation of the astragalus with the navicular is . . . diagnostic [for multituberculates], as it is an extreme and decidedly unique modification of the primitive mammalian trait of a convex, or sellar navicular AN facet.' We agree with Szalay and argue that the astragalonavicular articulation worked in multituberculates differently than in therians and all other mammals. In multituberculates the facet for the navicular is saddle-shaped, extending along the dorsal and medial sides of the base of the 'astragalar head', rather than being situated on its distal end as in therians.

It follows that multituberculates 'solved' the movements in the astragalonavicular joint differently than therians and monotremes: there is abduction–adduction in a horizontal plane, around a vertical axis, in multituberculates; pronation–supination (rotation around longitudinal axis) in therians; and flexion–extension in dorsoplantar direction, around a transversal axis, in monotremes (personal observations by PPG). The movements in the astragalonavicular joint in other mammals still need to be explored.

Functional anatomy

Reconstruction of locomotion

Introductory remarks

It was generally accepted after the analysis of the function of the multituberculate pes by Krause & Jenkins (1983) and Jenkins & Krause (1983) that at least some multituberculates were arboreal in their habits. Rowe & Greenwald (1987, p. 25A) went further and stated: 'It now appears more likely that arboreality was the ancestral habit of Theriiformes [the taxon comprising the immediate common ancestor of Multituberculata and Theria and all its descendants] and that the origin of Multituberculata involved dietary specialization within that niche.' We do not see any reason to presume that arboreality was primitive for multituberculates. We shall argue below that the Late Cretaceous Asian multituberculates studied by us were possibly terrestrial runners. For this reason the discussion on the reconstruction of their locomotion begins with a recapitulation of terrestrial gaits.

In symmetrical gaits a movement of one forelimb is followed by the movement of either of the hind limbs, after which the next forelimb and the next hind limb move; at the same time the lateral flexures of the body take place. As a result the pattern of limb movements on the right side of the body is a mirror image of those on the left. In asymmetrical gaits the forelimbs move first and then the hind limbs. There is no symmetry of movements between the right and left sides of the body, no lateral flexures, and the extension and flexion of the vertebral column occur in a sagittal plane (Howell 1944; Sukhanov 1974; Gambaryan 1967, 1974).

Amphibians, most extant reptiles, and monotremes have a sprawling limb posture, and the proximal segments of their limbs, when seen in caudal and cranial views, are arranged at about 90° to the parasagittal (vertical) plane. We refer to the sprawled limbs as abducted. The angle under which the abducted limbs are positioned may be much smaller than 90°. With limbs arranged at 90°, the humerus and femur show their full lengths in dorsal view, and if the angle is smaller the humerus and femur appear shorter in both dorsal and lateral views. The limbs of therian mammals were traditionally referred to as parasagittal. However, as demonstrated by Jenkins (1971b), *Didelphis, Tupaia, Rattus* and *Mustela* have humeri that function at angles of 10–30° from the parasagittal plane, while the femoral axis is positioned 20–50° from the parasagittal plane (see also Jenkins 1974 and Jenkins & Weijs 1979). Even in cursorial mammals the femur is slightly abducted (Jenkins & Camazine 1977). The gait of therians, however, differs from that of animals with sprawling posture. In therians the distal end of the femur moves in a parasagittal plane and during the propulsive phase no adduction takes place (Jenkins & Camazine 1977, Figs. 1–3). In monotremes (Pridmore 1985), during the propulsive phase adduction of the femur occurs, and as we shall argue below, the same apparently took place in multituberculates. Because of this difference (in spite of the more or less abducted position of the femur in therian mammals), in the discussion that follows, we conditionally accept that the limbs of therians move in a parasagittal plane, and we refer to them as 'parasagittal'.

In extant small mammals there occur three basic quadrupedal asymmetrical gaits: primitive ricochet (e.g., *Meriones*), ricochet (e.g., *Allactaga*) and gallop (e.g., *Lepus*) (Gambaryan 1967, 1974). In galloping mammals there are two phases of flight in the cycle: extended flight and gathered flight (see 'Terminology'). In ricochetal and primitively ricochetal mammals only the extended flight occurs. In galloping mammals (e.g., in *Lepus*) the angle of take-off is smaller than in the primitively ricochetal and ricochetal mammals (Gambaryan *et al.* 1978).

Function of the spinous processes of the lumbar vertebrae

The function of the spinous processes of the vertebrae in mammalian gait has been discussed by Slijper (1946), Brovar (1935, 1940) and Kummer, (1959a, 1959b) who demonstrated that the length of the spinous processes is related to the load on the flexion and extension of the spine. In order to establish the relative lengths of the spinous processes in multituberculates and in small extant therian mammals, we measured the lengths of the spinous processes in relation to the respective lengths between the anterior margin of the prezygapophysis and the posterior margin of the postzygapophysis (Fig. 4 and Table 2).

The partly broken-off spinous processes of the last lumbar vertebrae have been preserved in *Nemegtbaatar* (Figs. 8A, 11 and 35). In Fig. 36B$_2$ we give two reconstructions of the spinous processes in *Nemegtbaatar*, marked by dashed and dotted lines.

In extant mammals that use asymmetrical gaits, there is a correlation between the lengths of the spinous processes and the mass of m. erector spinae. If our first reconstruction of the lengths of the spinous processes in *Nemegtbaatar* (dashed lines in Fig. 36B$_2$) were true, one may conclude that the mass of erector spinae in *Nemegtbaatar* was relatively greater than in highly specialized ricochetal mammals such as, e.g., *Allactaga* (Table 3, Fig. 47). If our second reconstruction were true (dotted lines in Fig. 36B$_2$), then the mass of the erector spinae would be smaller than in *Allactaga*, but still greater than in, e.g., *Meriones*.

Table 2. Ratio of the length of the transverse and spinous processes of the last lumbar vertebrae to the length between the pre- and postzygapophyses (in percent).

Cat. No.	Species	L4 spi	L4 tr	L5 spi	L5 tr	L6 spi	L6 tr	L7 spi	L7 tr
81	*Nemegtbaatar gobiensis,* 1	94		100	106	100		100	133
81	*Nemegtbaatar gobiensis,* 2	81		86	106	86		86	133
13951	*Antechinus stuarti*	60	35	62	37	62	37	63	34
1380	*Elephantulus rozeti*					90	62	93	102
800	*Sciurus persicus*	41	44	47	54	55	61	72	67
175	*Citellus fulvus*	27	20	30	32	32	37	37	59
159	*Citellus xanthoprymnus*	26	27	28	37	30	41	35	53
59	*Spermophilopsis leptodactylus*	49	52	63	60	65	68	60	96
79	*Glis glis*	34	34	38	42	43	53	45	59
83	*Rattus norvegicus*	63	37	65	47	64	47	55	45
314	*Mesocricetus brandti*	27	22	30	28	30	37	50	48
311	*Cricetulus migratorius*	30	50	35	54	35	70	37	63
403	*Calomyscus bailwardi*	52	31	58	55	62	67	79	85
428	*Meriones tamariscinus*	75	87	79	97	90	110	80	111
114	*Meriones meridianus*	56	41	66	57	67	80	71	109
117	*Allactaga jaculus*	75	61	79	66	85	73	98	99
143	*Lepus europaeus*	70	125	73	126	74	131	70	121

Cat. No. (catalogue numbers) refer to ZPAL MgM-I/ collection for *Nemegtbaatar* and to ZIN, Laboratory of Mammals collection for all other taxa; spi = spinous process; tr = transverse process. Two different data for *Nemegtbaatar gobiensis* refer to two different reconstructions of the lengths of spinous processes in Fig. 36 B$_2$; 1 = the lengths marked by dashed lines; 2 = by dotted lines. For jumping indexes (jumping distance to body length) see Table 5.

As a rule, the spinous processes of the lumbar vertebrae in mammals are longer than those of the caudal vertebrae. In *Kryptobaatar* the spinous process of L6 and L7 (numbers 2 and 3, respectively, in Fig. 2) have been broken, while Cd1 (number 5) has been completely preserved. If the spinous process of L6 was at least as long as that of Cd1, it would be 91% of the length between the pre- and postzygapophysis, while that of L7 would be 97% of the same length. In *Chulsanbaatar* (ZPAL MgM-I/83) these processes are not well observable (Fig. 18). We measured the preserved parts of broken L4, L5 and L7, and they correspond to 71%, 65% and 54%, respectively, of the length between the corresponding pre- and postzygapophyses; this shows that even when broken they are relatively longer than the corresponding spinous processes in *Meriones meridianus* (Table 2). This suggests that our first (dashed lines) reconstruction of the lengths of the spinous processes in *Nemegtbaatar* (Fig. 36B$_2$) is more probable.

Table 3. Weight of erector spinae muscles in percent of the total weight of the skeleton and muscles, excluding hind limbs.

Species	a iliocos.	b semispi.	c longi.	$\sum$a+b+c = e. spinae	Length ratio
Mesocricetus raddei	0.10	0.59	1.18	1.27	30.0
Ellobius lutescens	0.51	0.41	1.3	2.22	34.0
Meriones blackleri	0.29	0.78	2.86	3.93	68.0
Rattus norvegicus	0.21	0.43	2.11	2.75	60.5
Allactaga elater	1.09	1.06	3.25	5.40	97.0
Nemegtbaatar gobiensis					100.0
					86.0

iliocos. = iliocostalis; semispi. = semispinalis dorsi; longi. = longissimus dorsi; e. spinae = erector spinae. Length ratio = ratio of the average length of the spinous processes of L6 and L7 to the distance between the prezygapophyses and postzygapophyses (in percent). Two numerals for *Nemegtbaatar* refer to two reconstructions of the lengths of the spinous processes in Fig. 36 B$_2$. The first numeral (100.0) corresponds to the dashed lines, the second (86.0) to the dotted lines.

We found in mammals that use asymmetrical locomotion a correlation between the length of the spinous processes of the last lumbar vertebrae and the length of the jump (Table 2 and Fig. 47). We define the jumping index (after Zug 1972) as a ratio of the jumping distance to the body length. The spinous processes of *Nemegtbaatar* (if the reconstruction in dashed lines were true) would be relatively longer than in all the taxa cited in Table 2, which indicates the capability for strong extension and flexion of the vertebral column in a sagittal plane and long asymmetrical jumps.

Function of the pelvic muscles

For the analysis of the locomotion it is important to reconstruct the topography of the pelvic muscles. We recon-

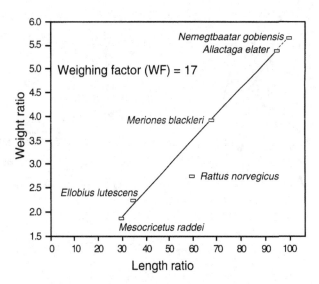

Fig. 47. Weighing factor, WF, in small modern mammals and *Nemegtbaatar*, representing a correlation between the mass of m. erector spinae and of the mean height of the two last spinous processes of the lumbar vertebrae. Numerals on abscissa (length ratio) represent mean length of spinous process of two last lumbar vertebrae (L$_s$) as percentage of distance between prezygapophysis and postzygapophysis (L$_z$). Numerals on ordinate (weight ratio) represent mass of erector spinae (W$_m$) as percentage of total mass of skeleton and muscles of body (W$_t$), excluding mass of hind limbs. The weighing factor WF can be expressed as length ratio/weight ratio. In measured extant mammals WF is 17, except in *Rattus norvegicus*, where it is 22. Assuming WF = 17 also for *Nemegtbaatar*, a tentative value for its weight ratio is 5.7. (See also Table 3.)

structed the origin of the pelvic muscles in *Nemegtbaatar*, the scars of which are discernible on the very well preserved bone surface (Fig. 37A), and we compare them with those of *Meriones tamariscinus* and *Antechinus stuarti* (Fig. 38), which are of approximately the same size as *Nemegtbaatar*. In these modern taxa all of the femoral adductors originate caudal to the vertical line extending ventrally from the acetabulum; in *Nemegtbaatar* they originated ventral to the acetabulum and cranial and caudal to the mentioned vertical line. The distance between the acetabulum and the ventral margin of the symphysis (although this is broken) is, in *Nemegtbaatar*, much greater than in *Meriones* and *Antechinus*. In *Kryptobaatar*, in which the ventral keel has been preserved, this distance is still relatively greater than in *Nemegtbaatar*.

We also suggest that the ventral keel of the multituberculate pelvis, a feature not found in other mammals, developed as a response to the origin of femoral adductors ventral to the acetabulum. The fenestration of the keel (postobturator foramen or notch) is possibly related to muscular attachments, as speculated by Krause & Jenkins (1983).

In order to understand how the hip joint worked in multituberculates, we compared it with that of *Rattus norvegicus*, the running mechanism of which has been studied by cinematography (Kuznetsov 1983). The configuration of

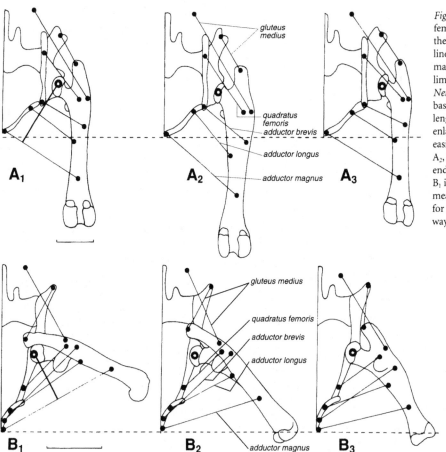

Fig. 48. Diagram illustrating position of femoral adductors and gluteus medius during the propulsive phase in caudal view. Dashed lines denote horizontal plane. □A. Eutherian mammal exemplified by *Rattus* ('parasagittal' limbs). □B. Multituberculate, exemplified by *Nemegtbaatar*, reconstructed partly on the basis of *Kryptobaatar* (abducted limbs). The lengths of the pelves of both taxa have been enlarged to the same size to make comparisons easier. A₁, B₁, beginning of propulsive phase; A₂, B₂, middle of propulsive phase; □A₃, B₃, end of propulsive phase. Bold lines in A₁ and B₁ illustrate how moment arm has been measured for adductor magnus. Moment arms for other muscles were measured in the same way. Scale bars 10 mm. (See also Table 4.)

the pelvic musculature in *Rattus* is generally the same as in *Meriones* and *Antechinus* (Fig. 38).

Figs. 48 and 49 illustrate a diagrammatical comparison of the geometry of the bones and muscles of the hip joint during the propulsive phase in caudal and lateral views in *Rattus* and *Nemegtbaatar* (based in part on *Kryptobaatar*). The moment arm of a muscle that medially retracts the femur is the distance between a point on the femoral head at the axis of rotation (fulcrum) and the muscle; the moment arm is perpendicular to the direction of contraction of the muscle fibers (bold lines in Figs. 48A₁, B₁ and 49A₁, B₁). We determined angles of the direction of muscle contraction between the pelvis and femur by measuring angles between the horizontal plane (perpendicular to the sagittal plane in caudal view) and the direction of the muscle contraction. Figs. 48 and 49 and Table 4 give the data on the moment arms of the retractors of the hip joint in *Rattus* and in *Nemegtbaatar*. In *Rattus* the retractors during the entire propulsive phase work to move the femur caudally, with the exception of adductor longus, which during the end of the phase (Fig. 49A₃) prevents further caudal movement of the femur; in this position adductor longus acts as a protractor of the femur. In *Nemegtbaatar*, during the beginning of the propulsive phase (Fig.

49B₁), adductor longus and add. magnus prevented the caudal movement of the femur. In *Nemegtbaatar* the muscle attachments of these two adductors are relatively large, which indicates their great mass.

The moment arm alone, without estimation of the force of a muscle, is relatively uninformative, but the ratio of adductor to retractor moments allows one to evaluate the main function of a given muscle. As can be seen in Table 4, all the adductors and quadratus femoris in *Rattus* have retractor moments greater than a sum of all adductor moments. In *Nemegtbaatar*, however, the adductor moments for all the adductors and quadratus femoris are greater than the retractor and protractor moments. This shows that in *Rattus* the main movements of the femur are due to retraction (in a craniocaudal direction). For this reason add. longus is relatively small, its mass being 7–8% of a sum of the mass of all the adductors of the femur and quadratus femoris (measured by us). In contrast, the main movements in *Nemegtbaatar* were due to the action of the adductors (in both medial and craniocaudal directions). In *Nemegtbaatar* the relatively great protractor moments of add. magnus and add. longus and the reduced retractor moments of add. brevis and quadratus femoris resulted in a more vertical shift

Fig. 49. Same taxa and skeletal parts as in Fig. 48, in lateral view. For explanations, see Fig. 48. Scale bars 10 mm.

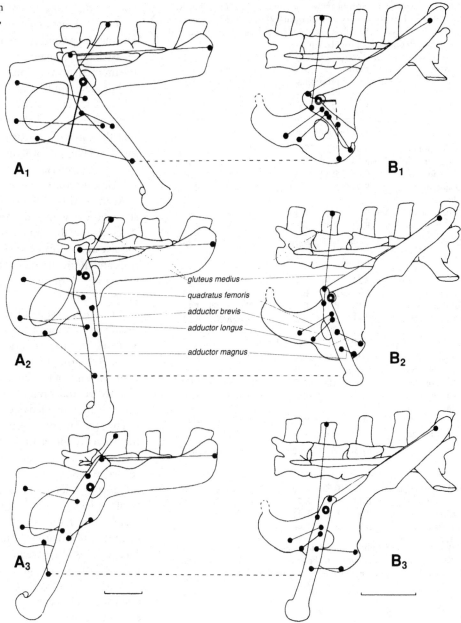

of the pelvis than in therian mammals (in which the protractor moment of add. longus is very small and add. magnus has only the retractor moment). The great retractor moments hinder the anterior excursion of the pelvis (and body). Therefore we speculate that in multituberculates the pelvis moved steeply vertically during the propulsive phase (as illustrated in Figs. 51–53) which resulted in a steep jump trajectory.

In extant, small, fast-running mammals the two phases of flight occur only in forms in which the angle of take-off is less than 7°. If the angle is greater, the hind limbs touch down before the forelimbs take off, and the phase of gathered flight does not occur (Gambaryan *et al.* 1978). If multituberculates

had a very steep trajectory of jump, they possibly did not develop the gathered phase.

The main difference in the mechanics of the propulsive phase in therians and multituberculates is that in therians retraction of the femur predominates (designated here retractoral mechanics), whereas in multituberculates adduction predominates (adductoral mechanics). For retractoral mechanics it is important that the adductoral component is balanced by the action of the gluteals. Therefore, in *Rattus norvegicus* the mass of m. gluteus medius is 1.3 times greater than the combined mass of all the adductors and quadratus femoris (Gambaryan, in preparation). This shows that gluteus medius, working as an abductor, balances the action of

Table 4. The moment arm of muscles at the hip joint (in millimeters, measured in Figs. 48 and 49 and enlarged ×2) and angle of muscular contraction between the pelvis and femur (in degrees) in *Rattus* and *Nemegtbataar* during the propulsive phase.

Muscles	*Rattus* Part of prop. phase			*Nemegtbaatar* Part of prop. phase		
	1	2	3	1	2	3
Moment arm: lateral view						
Adductor longus	R 16	R 2	P 13	R 1	P 16	P 21
Adductor brevis	R 23	R 26	R 25	R 8	R 8	R 7
Adductor mag.	R 35	R 37	R 31	R 11	P 23	P 31
Quadratus fem.	R 9	R 10	R 10	R 5	R 7	R 7
Gluteus medius						
pars sacralis	R 6	R 4	R 4	R 3	R 3	R 4
pars iliacus	R 14	R 15	R 15	R 6	R 6	R 4
Σ +/-	103/0	94/0	92/13	34/0	24/39	22/52
Angle						
Adductor longus	B 24	B 84	B 140	A 124	A 156	A 175
Adductor brevis	B 4	B 8	B 3	A 42	A 45	A 35
Adductor mag.	B 13	B 40	B 80	A 92	A 158	A 178
Quadratus fem.	B 14	B 16	B 12	A 36	A 28	A 22
Moment arm: caudal view						
Adductor longus	+19	+15	+20	+19	+19	+22
Adductor brevis	+11	+10	+13	+14	+12	+19
Adductor mag.	+37	+34	+35	+27	+31	+35
Quadratus fem.	-5	-5	-3	+10	+9	+10
Gluteus medius						
pars sacralis	-10	-11	-8	-19	-16	-17
pars iliacus	-15	-18	-15	+8	+8	-3
Σ +/-	67/30	59/34	68/26	76/19	79/16	86/20
Angle						
Adductor longus	B 35	B 50	B 33	A 38	A 28	A 19
Adductor brevis	B 49	B 38	B 25	A 44	A 42	A 34
Adductor mag.	B 30	B 40	B 31	A 28	A 16	A 14
Quadratus fem.	B 50	B 53	B 46	A 45	A 42	A 35

Parts of the propulsive phase: 1 beginning; 2 middle; 3 final. + = adduction; − = abduction. A = above horizontal plane; B = below horizontal plane; R = retraction; P = protraction; mag. = magnus; fem. = femoris.

all the adductors and quadratus femoris; at the same time all these muscles act as retractors of the femur (Table 4).

For adductoral mechanics it is important to reduce the retractor component, and therefore this is partly balanced by the protractors. The complete reduction of the retractor component is, however, impossible, as in such a case the anterior propulsion would disappear and the jump would be vertical. In multituberculates, in order to retain the retractor component, the femur rotated and therefore gluteus medius apparently acted differently in *Nemegtbaatar* than in *Rattus*. In *Nemegtbaatar* the lesser trochanter is very prominent, situated in the middle of the ventral surface of the femur, while in *Rattus* it is much smaller, placed at the medial margin of the femur. We infer that in multituberculates, m. iliopsoas rotated the femur to such extent that its dorsal surface at the beginning of the propulsive phase faced later-

ally. At that time the moment arm of iliopsoas was the greatest and that of gluteus medius the smallest (Fig. 50A). During the propulsive phase the rotation of the femur increased, the moment arm of gluteus medius increased and the moment arm of iliopsoas decreased (Fig. 50B, C). At the end of the propulsive phase the dorsal surface of the femur was directed cranially, as in modern therian mammals. For the pronation (medial rotation) of the femur it is desirable that the force of gluteus medius would act parallel to the sagittal plane. The prolongation of the ilium and the reflection of its cranial ends in *Nemegtbaatar* (seen in *Kryptobaatar*, Fig. 3A) contribute to this aim.

At the end of the propulsive phase, m. gluteus medius started to act as an abductor and retractor of the femur. We speculate that this caused in multituberculates an increase in the size of the greater trochanter, elongation of the femoral neck and the medial shift of the head, and at the same time an elongation of the ilium for the insertion of the powerful gluteus medius.

We also suggest that the dorsocaudally open (referred to usually as dorsally open) structure of the multituberculate acetabulum, may be due to the action of the retractor muscles. During the vertical movement of the pelvis, the main pressure of the femoral head in multituberculates was directed ventrally rather than dorsally, and that is why the acetabulum remained open dorsocaudally.

The acetabulum is open dorsally in some modern therian mammals, as for example in *Petaurista* (Elftman 1929) and *Desmana* (Gasc & Gambaryan, in preparation), which is related to gliding and swimming habits, respectively. In both these types of locomotion, the femur is abducted at almost right angles to the sagittal plane, which results in an open acetabulum. Elftman (1929) stated that the acetabulum is open dorsally in arboreal marsupials such as *Didelphis* and *Pseudochirus*. We have not had an opportunity to examine the skeletons of *Pseudochirus*, but in several specimens of *Didelphis* examined by us we have not found the dorsally open acetabulum. We also examined skeletons of two other arboreal marsupials, *Trichosurus velpecula* and *Phalanger maculatus*, and in both of them we found the acetabulum dorsally closed.

Simpson & Elftman (1929) reconstructed the movements of the femur in ?*Eucosmodon* sp. and noted its rotation and abducted position. Jenkins & Krause (1983) argued that in *Ptilodus* the femur was abducted about 45° from the sagittal plane. We speculate that the multituberculate femur was abducted from 60° at the beginning of the propulsive phase to about 30° at the end. Abduction greater than 60° was physically impossible, as in such a position the greater trochanter would interfere with the ilium (Fig. 48B₁). During the beginning of the propulsive phase, the adduction of the femur displaced the pelvis strongly craniodorsally (Figs. 51–53). The range of pelvic movement decreases exponentially with the decrease of the angle of abduction, as it depends on the cosine of the angle (Figs. 48 and 51). An abduction of less

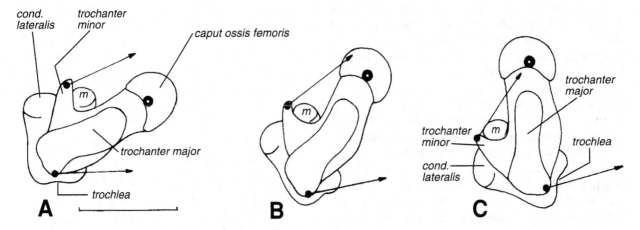

Fig. 50. Nemegtbaatar gobiensis (ZPAL MgM-I/81). Diagram illustrating inferred rotation of right femur around longitudinal axis during beginning (A), middle (B) and end (C) of the propulsive phase. Femur is seen in proximal view. In A the ventral side is up; in B and C it is moved to the left. Black circle on femoral head denotes center of rotation; m = medial condyle (condylus medialis). Upper arrow denotes direction of constriction of m. iliopsoas, lower arrow of m. gluteus medius and m. gluteus profundus. Scale bar 5 mm.

than 30° would move the pelvis only insignificantly upwards and would not result in an increase of the thrust of propulsion. On this basis we suggest that the femur possibly did not abduct less than 30° in multituberculates.

The main difference between the hind-limb movements of multituberculates and those of modern therians is that in multituberculates adduction of the femur took place during the whole propulsive phase (Fig. 51). In therians there is no adduction of the femur during the propulsive phase (Jenkins & Camazine 1977). Whereas in therians the distal end of the femur moves only in a parasagittal plane during the propulsive phase, in multituberculates it moved in both parasagittal and transverse planes.

Figs. 51–53 present the inferred movements of the hind limbs and pelvis in multituberculates (in *Nemegtbaatar*, based in part on *Kryptobaatar*) during the propulsive phase in caudal, lateral and dorsal views. At the beginning of the propulsive phase the femur (as argued above) lay about 60° to the parasagittal plane and the pes stroke the ground with all the phalanges. During propulsion the ungual phalanges remained at the same place, but at the end the first and fifth digits were clear of the ground. During propulsion the femur adducted and rotateed cranially about its longitudinal axis (upper arrows in Figs. 51 and 53, curved arrow in Fig. 52); the crus rotated in the knee joint and the calcaneal tubercle (and tarsus and metatarsals) rotated laterally (lower arrows in Figs. 51 and 53). As a result of the hind limbs movements and rotation, the pelvis was shifted craniodorsally.

Function of the tail

In modern small therian mammals, during the propulsive phase the impulse of the force of the limbs is directed craniodorsally. If this impulse extends cranial to the center of gravity, the tail at the end of the propulsive phase moves ventrally, as demonstrated by Fokin (1978, Figs. 6, 9, 11, 15) for many rodents. However, in some modern mammals, the impulse of the force of limbs extends caudal to the center of the gravity. In such forms, (e.g., *Pygerethmus*; see Fokin 1978, Fig. 10) at the end of the propulsive phase the tail moves upwards, in order to balance the ventral rotation of the anterior part of the body. We speculated under 'Function of the pelvic muscles' that multituberculates had a steep trajectory of jumps. If so, the impulse of the force of limbs was directed craniodorsally, caudal to the center of gravity. Therefore we believe that before the take-off, the multituberculate tail had to move upwards (Fig. 52), as in *Pygerethmus*.

Aristov *et al.* (1980) demonstrated that a long, heavy tail commonly occurs in small terrestrial mammals and acts as a balancing organ not only during jumping, but also during turning. The tail is also used as a prehensile organ in many arboreal mammals.

Krause & Jenkins (1983) regarded the tail of the Paleocene multituberculate *Ptilodus kummae* to be prehensile and cited four characters that they considered characteristic for prehensile-tailed mammals: great length of the tail; haemal arches developed along nearly the entire length of the tail; robust transverse processes; and large sacral spinous processes, nearly equalling in height the spinous processes of the posterior lumbar vertebrae. All these characters occur in arboreal mammals, but at the same time they are found in terrestrial running and jumping mammals. We have examined the skeletons of *Macropus rufus*, *Petrogale penicillata*, *Elephantulus rozeti*, all the species of *Allactaga*, all the species of all the genera of the Dipodinae, and many others, and we have found these four characters in all of them. The only feature that characterizes prehensility and is absent in terrestrial forms is the corn (a local hardening and thickening of the skin, formed especially on the toes) on the ventral side of the tail, which is not preserved in fossils. In modern mammals the corn on the tail occurs, for example, in *Arctictis binturong* (Viverridae) and *Brachyteles arachnoides* (Cebidae). In these and many other taxa with corn, the vertebrae in the area of corn widen and then at the very end of the tail they narrow again. This is the only character that allows one to define the prehensile tail in fossil mammals, which, however, is not known in multituberculates (including *Ptilodus*, see Krause & Jenkins 1983, Table 2). The few caudal vertebrae preserved in *Kryptobaatar* and *Catopsbaatar* and the structure of the lumbar vertebrae in *Nemegtbaatar* (see 'Osteological descriptions') allow us to assume that the tail in the studied Asian multituberculates was long and similarly built as in *Ptilodus*.

Function of the transverse processes

In *Nemegtbaatar* the transverse processes of the lumbar vertebrae are robust (Figs. 11, 35, 36B); their lengths relative to the distances between the prezygapophyses and postzygapophyses are greater than in most taxa cited in Table 2 and smaller only than those of *Lepus europaeus*, which belongs to a different size category.

With 'parasagittal' position, the limbs strike the substrate close to the sagittal plane. Because of this the force rotating the pelvis, sacrum and the lumbar vertebrae around the longitudinal axis is relatively small (Jenkins & Camazine 1977, Table I). With abducted position this force increases. In *Nemegtbaatar* the long transverse processes are possibly an adaptation for balancing the force of the vertebral rotation during the propulsive phase. In *?Eucosmodon* sp. and *Ptilodus kummae* the spinous and transverse processes were approximately as long as in *Nemegtbaatar* (Krause & Jenkins 1983, Figs. 1, 2, 7 and 28).

Abducted limbs occur in amphibians and reptiles, which employ symmetrical locomotion with strong lateral flexures of the body. In relation to this the transverse processes of the lumbar vertebrae are short, as their elongation would hinder the lateral flexures. The long transverse processes of the lumbar vertebrae in multituberculates thus indicate asymmetrical locomotion.

Proportions of hind limb segments

As may be seen in Table 5, the tibia in *Kryptobaatar* is relatively shorter than in slow runners such as the Cricetidae (*Mesocricetus raddei*) and terrestrial Sciuridae (*Citellus xanthoprymnus* and *C. fulvus*). Mt III in *Kryptobaatar* (compared to the length of the femur) is relatively longer than in *Mesocricetus raddei* and *Cricetulus migratorius*, and similar to that in the terrestrial Sciuridae, but shorter than in taxa more specialized for jumping (Table 5). Only the segments of the

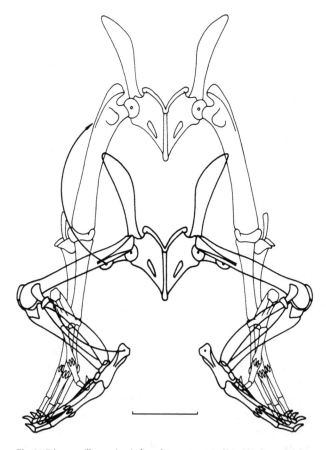

Fig. 51. Diagram illustrating inferred movements of hind limbs and pelvis in *Nemegtbaatar* (based in part on *Kryptobaatar*), at beginning (bold lines) and end of propulsive phase, in caudal view. Upper arrow denotes cranial rotation of femur about its longitudinal axis; lower arrow denotes lateral rotation of tarsus. See text for explanation. Scale bar 10 mm.

third digit in the pes of *Kryptobaatar* are relatively longer than in extant mammals cited in Table 5. Relatively short tibia and metatarsals would indicate that *Kryptobaatar* was running rather slow, similar to, for example, *Cricetulus migratorius* (4 m per second; unpublished experimental data of Gambaryan).

However, there is a contradiction between the data obtained from the lengths of the spinous processes of the lumbar vertebrae (which indicate long jumps in studied multituberculates) and those obtained from relative lengths of the hind-limb segments (which indicate relatively short jumps). This contradiction may be the result of the abducted position of multituberculate limbs, because of which their movements were different from mammals with 'parasagittal' limbs. The direct transposition of data obtained from the analysis of gaits in modern mammals would thus be incorrect.

In order to understand the mechanics of multituberculate locomotion, we made an attempt to compare it with that of frogs, which are the only modern vertebrates that use asym-

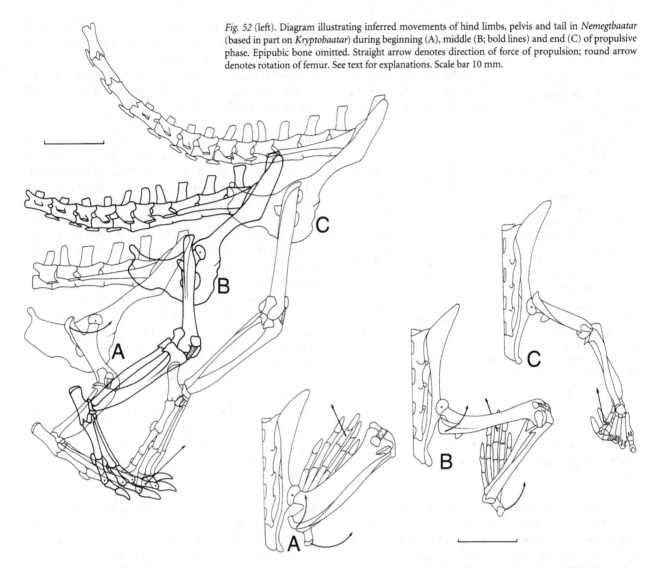

Fig. 52 (left). Diagram illustrating inferred movements of hind limbs, pelvis and tail in *Nemegtbaatar* (based in part on *Kryptobaatar*) during beginning (A), middle (B; bold lines) and end (C) of propulsive phase. Epipubic bone omitted. Straight arrow denotes direction of force of propulsion; round arrow denotes rotation of femur. See text for explanations. Scale bar 10 mm.

Fig. 53 (right). Diagram illustrating inferred movements of hind limbs, pelvis and sacrum in *Nemegtbaatar* (based in part on *Kryptobaatar*) during beginning (A), middle (B) and end (C) of propulsive phase, in dorsal view. Straight arrows denote direction of force of propulsion; left round arrows denote rotation of femur, round right arrows denote rotation of calcaneal tubercle. See text for explanation. Scale bar 10 mm.

metrical jumps with abducted limbs. In frogs the trajectory of jump is very steep: the angle of take-off is 30–35° (Gans 1961, 1974) or even 45° (Gray 1968). We speculate that in multituberculates during the jump the hind limbs probably moved rapidly medially (as in frogs), which resulted in a jump trajectory that was higher than in modern therian mammals. In small mammals with 'parasagittal' limbs studied by us, after the propulsive phase the center of gravity takes off at 3–10° to the horizontal plane (Gambaryan 1974; Gambaryan *et al.* 1978), and the trajectory of their jump is very low.

Casamiquela (1964; see also Leonardi 1987) described the tracks of a small mammal designated *Ameghinichnus patagonicus* from the Late Jurassic of Patagonia. He attributed to *Ameghinichnus* three types of tracks, one of which is sym-

metrical and two others asymmetrical, the asymmetrical ones were classified by him as gallop ('andar galopado'). *Ameghinichnus* had the hind limbs more abducted than the forelimbs and was digitigrade. At the beginning of gallop it placed the forelimbs slightly to the rear and inside the hind limbs. In the second phase of gallop all four feet were placed approximately at the same transverse line. The tracks are roughly reminiscent of those of *Microtus gud* and *Alticola strelzovi* figured by Gambaryan *et al.* (1978). Casamiquela attributed *Ameghinichnus* tentatively to Pantotheria. Although the evidence is inconclusive, we believe that it might belong to multituberculates which we regard as digitigrade (see below). If so, these tracks would indicate that multituberculates employed both symmetrical and asymmetrical gaits with jumps.

Table 5. Lengths of the hind limb segments in percent of the femur length, and the jumping index.

Cat. No.	Species	Tibia	Mt III	DIII Ph1	DIII Ph2	Ind	Hab
41	*Kryptobaatar dashzevegi*	84	33	22	16		t
1110	*Marmosa* sp.	121	25	19	12	?	s
1380	*Elephantulus rozeti*	140	72	14	9	11	r
46	*Sciurus persicus*	114	36	18	14	6–7	s
175	*Citellus fulvus*	98	32	17	11	2–3	t
159	*Citellus xanthoprymnus*	100	30	19	10	3–4	t
59	*Spermophilopsis leptodactylus*	108	34	19	13	6–7	t
79	*Glis glis*	107	34	21	13	5–6	s
83	*Rattus norvegicus*	113	43	18	11	4–5	t
11	*Mesocricetus raddei*	91	21	11	9	1–2	t
314	*Mesocricetus brandti*	92	19	10	6	1–2	t
311	*Cricetulus migratorius*	104	29	13	10	4–5	t
403	*Calomyscus bailwardi*	129	57	20	13	6–7	t
428	*Meriones tamariscinus*	116	46	18	12	6–7	t
114	*Meriones meridianus*	131	53	21	12	7–8	t
117	*Allactaga jaculus*	139	93	26	15	14–15	r

Cat. No. (catalogue number) refers to ZPAL MgM-I/ collection for *Kryptobataar* and to ZIN, Laboratory of Mammals collection for extant taxa. D = digit; Hab = habits; Ind = jumping index (ratio of jumping distance to body length); Mt = metatarsal; Ph = phalanx; r = ricochetal; s = scansorial; t = terrestrial. Jumping indexes are given after experimental unpublished data of Gambaryan (see Gambaryan *et al.* 1978 for description of methods).

Structure and function of multituberculate pes

In cursorial therian mammals the pes as a rule is situated parallel both to the sagittal plane and to the direction of movement. In animals with abducted limbs the pes may be placed in various positions. For example, in *Ornithorhynchus* the manus is directed almost parallel to the direction of movement, and the pes lies at an angle of about 70° (Pridmore 1984, Fig. 1). In *Tachyglossus* the manus is directed medially, at about 60°, and the pes is directed laterally at about 70° (Jenkins 1970a, Fig. 2). In lacertilians the manus and pes may be placed from transverse to almost parallel positions to the direction of movement (Sukhanov 1974; Leonardi 1987).

The first multituberculate pes ever found was that of ?*Eucosmodon* sp. (AMNH 16325) from the Early Paleocene of New Mexico, reconstructed by Granger & Simpson (1929). In their reconstruction (Fig. 23) Mt III is arranged parallel to the longitudinal axis of the tuber calcanei, and the distal part of the calcaneum is 'hanging in the air' and is not supported distally by any tarsal or metatarsal bone.

The skeleton of ?*Eucosmodon* sp. was subsequently studied by Krause & Jenkins (1983) and Jenkins & Krause (1983). They estimated that the longitudinal axis of the foot (passing along the third ray) deviated 30–40° from a sagittal plane. They demonstrated a wide range of the pedal mobility (especially abduction) in ?*Eucosmodon* and *Ptilodus*, characteristic for arboreal mammals that descend trees headfirst.

Szalay (1993) studied the tarsus of ?*Eucosmodon* and isolated calcanea and astragali of a ptilodontoid from the Gryde Locality of the Frenchman Formation, Alberta, and reconstructed relevant tarsal joints. His reconstruction of ?*Eucosmodon* pes is similar to those of Granger & Simpson (1929) and Krause & Jenkins (1983).

We reconstruct the pes of *Kryptobaatar* (Fig. 54) primarily on the basis of ZPAL MgM-I/41 (Figs. 2, 6A–C, 7) in which both pedes have been preserved in articulation, and on the basis of an articulated left pes of *Chulsanbaatar*, ZPAL MgM-I/99b (Fig. 25). We also used in this reconstruction unidentified multituberculate hind-limb elements from the Late Cretaceous of North America (Fig. 55), on which the articular surfaces are well preserved. We present a new reconstruction of the pes of ?*Eucosmodon* on the basis of AMNH 16325 (Figs. 56, 57). Although pedes of *Kryptobaatar* and ?*Eucosmodon* differ from one another in details, they share features that we regard as characteristic of multituberculates as a whole. In our reconstruction the astragalus is placed obliquely lateroproximally to mediodistally with respect to the tuber calcanei, Mt III is abducted about 30° from the longitudinal axis of the tuber calcanei, and Mt V articulates with the distal margin of the calcaneum medial to the peroneal groove.

In the three pedes of Mongolian multituberculates mentioned above, the distal part of the tarsus is preserved in its original position in articulation with the metatarsals. Our observations on the movements at the astragalonavicular joint in multituberculates agree with those of Szalay (1993). There is an extensive concave facet on the navicular, for articulation with the astragalus. In Mongolian multituberculates the astragalus is directed obliquely, but in *Kryptobaatar* it has been slightly shifted from its original position on both sides. In spite of the shifting it is obvious that the saddle-shaped facet on the mediodistal side of the astragalus (at the base of the 'astragalar head' of other authors), which extends below the short margin (medial in the reconstruction of ?*Eucosmodon* by Granger & Simpson 1929, and mediodistal in our reconstructions of *Kryptobaatar* and ?*Eucosmodon* – Figs. 54, 57), articulated with the concave facet on the proximal margin of the navicular. In an isolated unidentified multituberculate astragalus from the Late Cretaceous of North America (Fig. 55B–D) and in ?*Eucosmodon* sp. (Fig. 56C, D), the astragalonavicular facet is similarly extensive as in *Kryptobaatar* and extends from the mediodistal astragalar process medially around the corner below the shorter margin of the astragalus. If one would place the astragalus transversely, only a part of the astragalonavicular facet would match the concave facet on the navicular.

We made an attempt to place all of the tarsal bones of ?*Eucosmodon* sp. (AMNH 16325) in various positions, so that all of the articular facets of all the bones would fit one another, and only in the position shown in Fig. 57 did we obtain a perfect match. In this position the surface of the calcaneum bearing sustentacular and astragalocalcaneal facets, and usually referred to as dorsal (e.g., Krause & Jenkins 1983, Fig. 24; Szalay, 1993, Fig. 9.10), is placed obliquely medially, and the tuber calcanei is shifted in a plantar direction relative to the metatarsals (Fig. 57C, D).

In *Kryptobaatar* the proximal end of Mt V protrudes proximally, and there is a small facet on it that fits the distal margin of the calcaneum medial to the peroneal groove. A small medioproximal facet on Mt V articulates with the distal part of the lateral margin on the cuboid. In ?*Eucosmodon* on the plantar side of the distal margin of the calcaneum there is a facet for articulation with Mt V. On the proximal margin of Mt V, on the dorsal side, there is a distinct oblique facet (seen in Fig. 23 of Granger & Simpson 1929 and in Fig. 24 of Krause & Jenkins 1983) for articulation with the distal end of the calcaneum.

In all of the multituberculate calcanea the prominent laterodistal peroneal tubercle is separated from the calcaneal body by a deep groove (e.g., Figs. 55F–I, 56J). In some marsupials and rodents along a groove on the lateral surface of the calcaneum there extends the tendon of m. peroneus longus; This tendon passes along the notch on Mt V to the

groove on the cuboid and then extends to Mt I. In *Tachyglossus* the tendon of m. peroneus longus, as stated by Lewis (1963, p. 57), 'enters the sole through an aperture between the calcaneus and the cuboid. After establishing a slender attachment to the base of the fifth metatarsal, the major part of the tendon crosses the sole obliquely, as in other mammals, to be inserted on the flattened and disc-like first metatarsal.'

A peroneal groove on the multituberculate calcaneum can only be interpreted as housing the tendon of peroneus longus, as correctly recognized by Szalay (1993). However, in his reconstruction (Szalay 1993, Fig. 9.10), the peroneal groove points laterodistally rather than towards the bone housing the distal part of this tendon. In modern mammals, the bones of the pes with grooves for the tendon of peroneus longus contact one another (see, e.g., Lewis 1963, 1989; Evans & Christensen 1979; Schaller 1992; and others). We reconstruct the course of this tendon in ?*Eucosmodon* (Fig. 57) as passing along the bones that contact one another: extending along the dorsal side of the calcaneum and passing through the peroneal groove onto the plantar side to Mt V,

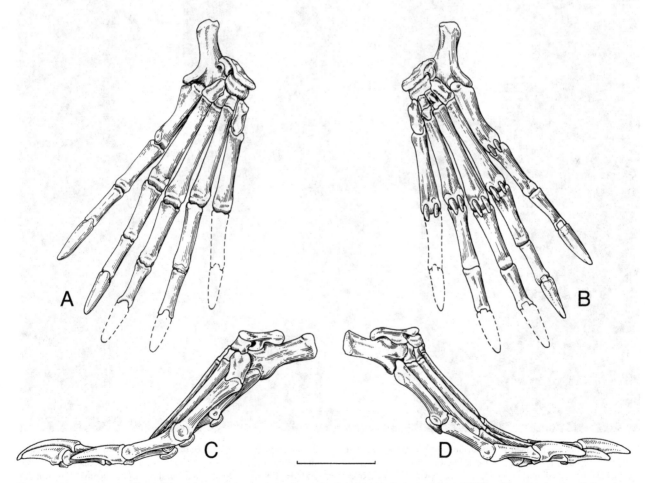

Fig. 54. Reconstruction of right pes of *Kryptobaatar dashzevegi* (ZPAL MgM-I/41), in dorsal (A), plantar (B), medial (C) and lateral (D) views. Phalanges not preserved are shown in A and B by dashed lines. Sesamoid bones not preserved in Mt I have been reconstructed. In A and B metatarsals and phalanges are drawn in the same plane. Scale bar 5 mm.

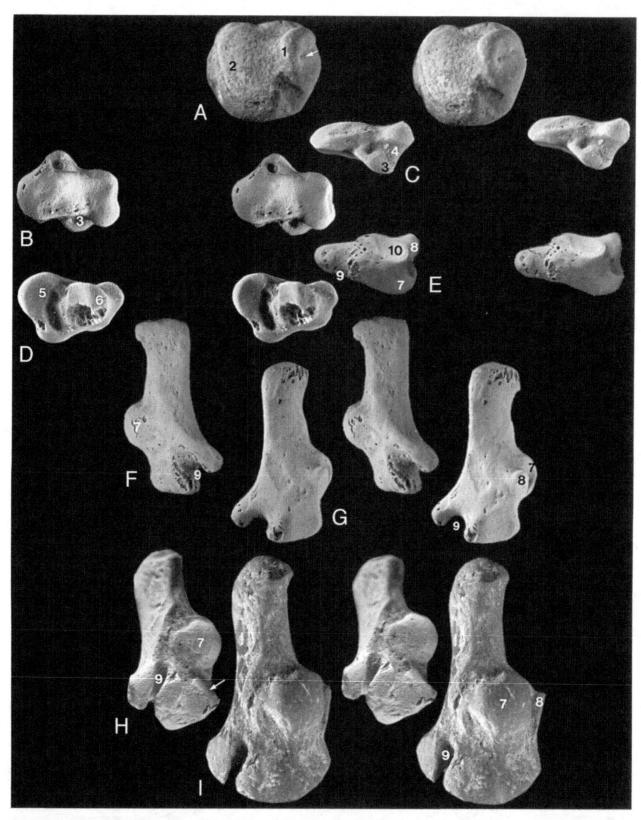

Fig. 55. Unidentified multituberculate postcranial elements from the Late Cretaceous of North America. □A. Left tibia, distal view (AMNH 117143). □B, C, D. Right astragalus (AMNH 118887), distal, dorsal and plantar views. In B the medioproximal margin is up; in D it is down. □E, F, G. Left calcaneum (ZPAL Z.p. M/123), distal, medioplantar and dorsolateral views. □H, I. Right calcaneum (AMNH 117990), laterodistal and dorsal views. Calcaneum in H is arranged at 45° in respect to position in I and therefore appears smaller. A is from the Lull II locality UCMP-V5620, Wyoming, all others from the Bug Creek Anthills site, Hell Creek Formation, Montana. 1 = medial malleolus; 2 = lateral condyle; 3 = mediodistal process of astragalus ('astragalar head'); 4 = saddle-shaped facet on astragalus for navicular; 5 = astragalocalcaneal facet; 6 = sustentacular facet of astragalus; 7 = astragalocalcaneal facet; 8 = sustentacular facet of calcaneum; 9 = peroneal groove; 10 = cuboid facet; arrow in A = pit for lig. collaterale mediale; arrow in H = cuboid facet. All ×6; stereo-pairs coated with ammonium chloride.

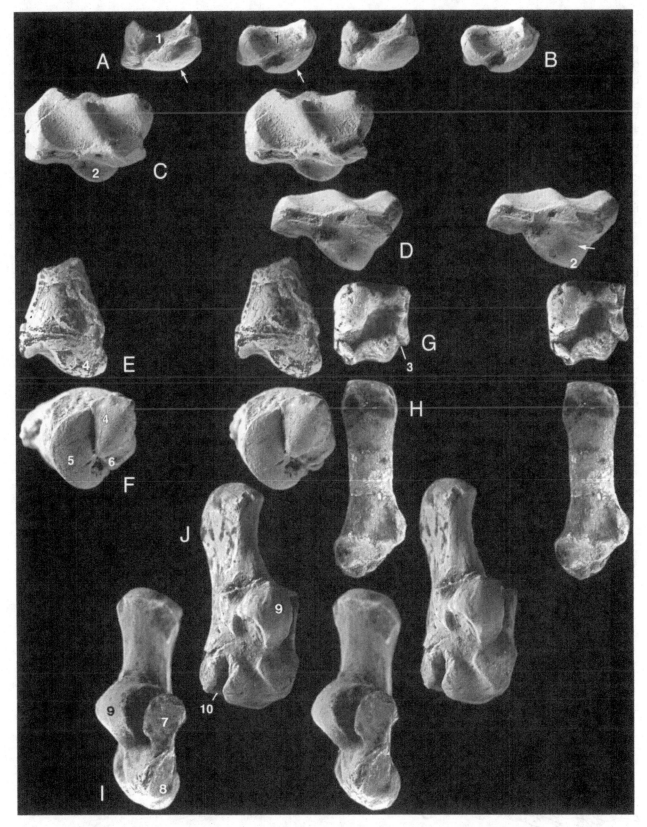

Fig. 56. Isolated right bones of ?*Eucosmodon* sp. (AMNH 16325), Nacimiento Formation, Early Paleocene, locality east of Kimbetoh, San Juan Basin, New Mexico. □A, B. Navicular in plantar and proximal views. □C. D. Astragalus in dorso–laterodistal and laterodistal views. □E, F. Distal part of the tibia in craniomedial and distal views. □G. Cuboid in lateral view, proximal margin is to the left. □H–J. Calcaneum in lateral (slightly distal), medial (slightly distal) and dorsal views. 1 = facet for astragalus in navicular; 2 = mediodistal process of astragalus ('astragalar head'); 3 = groove for tendon of m. peroneus longus in cuboid; 4 = medial malleolus; 5 = lateral condyle; 6 = medial condyle; 7 = sustentacular facet; 8 = cuboid facet; 9 = astragalocalcaneal facet; 10 = peroneal groove; arrows in A and B = distoplantar facet for articulation with ectocuneiform and mesocuneiform; arrow in D = saddle-shaped facet on astragalus for navicular. All ×4; stereo-pairs, coated with ammonium chloride.

then along the grooves on the plantar side of the cuboid and ectocuneiform onto Mt II, distal to the small tuber on the plantar side of Mt II, and finally reaching Mt I (see also details of individual bones in Figs. 55 and 56). If this were not the case, the presence of the peroneal groove would be a puzzle.

In the right astragalus (AMNH 118887) from the Hell Creek Formation of Montana, on the plantar surface there are two facets (Figs. 55D, 58). The astragalocalcaneal facet (lateral) is semilunar, as found by Krause & Jenkins (1983) in *Ptilodus kummae*. The sustentacular facet (medial) is almost entirely circular. If one would move the astragalus in a horizontal plane around the midpoint of the sustentacular facet, 10–15° each way, then on the middle part of the astragalocalcaneal facet one would produce a trace that corresponds almost exactly to the astragalocalcaneal (lateral) facet of the calcaneum (as seen in a dorsal horizontal projection) (Fig. 58).

The medial malleolus of multituberculate tibia fits the craniomedial side of the astragalus. On the medial surface of

the malleolus, as seen in an unidentified multituberculate tibia (arrow in Fig. 55A), there is a deep pit for the ligamentum collaterale mediale. This pit is also preserved, but less clear, on the tibia of *Nemegtbaatar* (arrow in Fig. 17G). Ligamentum collaterale mediale inserted on the astragalus. Krause & Jenkins, (1983, Fig. 30) described the rotation in the tibioastragalar and astragalocalcaneal joints in *?Eucosmodon* sp. We measured the rotation in these joints in *?Eucosmodon* sp. and found that the rotation in the tibioastragalar joint (about 30°, around the longitudinal axis of the tibia) was possibly more extensive than in the astragalocalcaneal joint (about 25°, around the same axis), the shared rotation being about 55°. In recent eutherian mammals, in contrast to multituberculates, rotation at the tibioastragalar joint is not possible because of the trochlear structure of the astragalus (Szalay 1984).

In some marsupials, however, e.g., *Didelphis marsupialis*, *Phalanger maculatus*, *Trichosurus vulpecula*, *Antechinus stuarti* and *Marmosa* sp. (personal observations by PPG), in

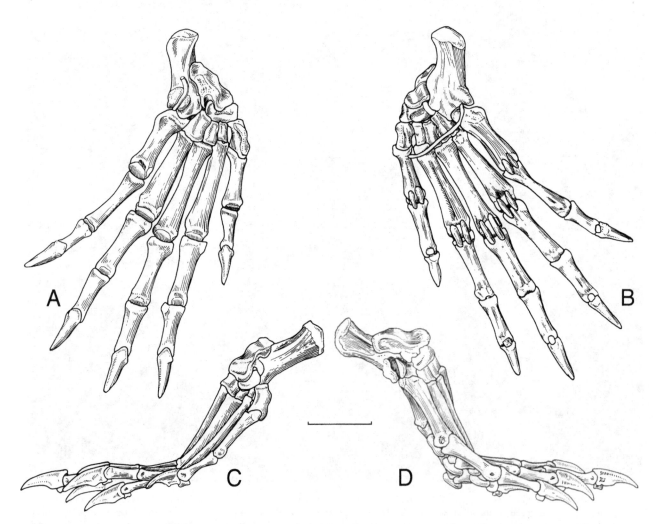

Fig. 57. Reconstruction of right pes of *?Eucosmodon* sp. (AMNH 16325), dorsal (A), plantar (B), medial (C) and lateral (D) views. Note the tendon of m. peroneus longus in A and B. Reconstruction of all digits is tentative, based on isolated phalanges. The pes is reconstructed with the adducted position of the first digit. In A and B metatarsals and phalanges are drawn in the same plane. Sesamoid bones are reconstructed. Scale bar 10 mm.

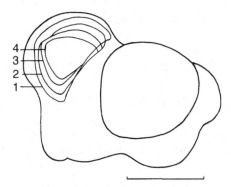

Fig. 58. Diagram based on an unidentified multituberculate astragalus (AMNH 118887) showing different positions of astragalocalcaneal facet during movement of astragalus on calcaneum. Plantar view. 1 = original position; 2 = 5° of the original position, 3 = 10°, 4 = 15°. Scale bar 2 mm.

addition to the rotation (pronation–supination as in eutherians) between astragalar head and navicular, there is also a rotation in the tibioastragalar joint (abduction–adduction) as in multituberculates. This rotation is possible because of the lack of the trochlea and presence of the flat tubercle on the dorsoproximal part of the astragalus. The tibial medial malleolus may rotate on the surface of this tubercle, around the longitudinal axis of the tibia.

We measured also the rotation in the astragalonavicular joint in *?Eucosmodon* (30°, abduction–adduction) and in the joint between the cuboid and the calcaneum (35°, flexion–extension). On the laterodistal end of the navicular there occurs a flat facet for articulation with the matching facet on the cuboid. These facets are of the same size and are flat, which may prevent movements at this joint. However, on the ectocuneiform and mesocuneiform there are proximal facets for articulation with a large distoplantar facet on the navicular (arrows in Fig. 56A, B). This suggests that the flat facets between the cuboid and navicular could separate during movements, as otherwise the movements between the navicular and both ectocuneiform and mesocuneiform would not be possible.

In *Kryptobaatar* (ZPAL MgM-I/41) on both sides, the calcaneum has been displaced in a plantar direction (Figs. 2, 6A, B, 7A–D). If one would place the calcaneum in its original position, as reconstructed in Fig. 54, the load of the calcaneum on the cuboid would extend perpendicular to the direction of movement of the pes (as it does in other mammals), rather than obliquely medially, as would appear from the reconstructions of Granger & Simpson (1929) and Krause & Jenkins (1983). An abducted position of the multituberculate pes and the compulsion to retain the load perpendicular to the direction of the movement explain the presence of an oblique, rather than distal (as in all other mammals) cuboid facet on the multituberculate calcaneum.

In modern reptiles, m. gastrocnemius inserts on the plantar aponeurosis (Romer & Parsons 1986). As argued by Gambarjan (1990), this is an adaptation for symmetrical

locomotion, in which the gastrocnemius constricts during the whole duration of the propulsive phase. Jenkins (1971a) described in cynodonts an incipient tuber calcanei, which indicates that the plantar aponeurosis possibly inserted partly on the calcaneum. In symmetrical locomotion the load on the tarsal joint is relatively small, as it is responsible only for retaining the speed of the gait. With asymmetrical jumps the pressure on the tarsal joint increases, and the tuber calcanei (which acts now as a lever for gastrocnemius) enlarges. At the same time the proximal part of the pes raises and becomes digitigrade. This confirms the conclusion of Kuznetsov (1985) that digitigrady is an early mammalian adaptation. We reconstruct the multituberculate pes as digitigrade (contra Simpson & Elftman 1928).

Pedal adaptations of Asian and North American multituberculates

The pedes of Asian taxa studied here differ in many details from those of the North American genera *Ptilodus* and *?Eucosmodon* described by Krause & Jenkins (1983). In the pes of *?Eucosmodon* the grooves on the dorsal surface of the astragalus are more prominent than in *Kryptobaatar*. The calcaneum in *?Eucosmodon* is relatively narrower, the astragalocalcaneal facet is more prominent (bulbous) and placed closer to the sustentacular facet, the peroneal tubercle is less shifted laterally, and the peroneal groove is narrower.

We were not able to measure the mobility in all the joints of the pes in Asian taxa studied by us. It seems, however, that the mobility at the astragalocalcaneal, astragalonavicular and tibioastragalar joints was smaller in *Kryptobaatar* and *Chulsanbaatar* than in *?Eucosmodon*. We base this conclusion on the structure of *Kryptobaatar* and *Chulsanbaatar* calcanea, in which the sustentacular and astragalocalcaneal facets, although poorly preserved, are possibly less prominent and more distant from one another than in *?Eucosmodon*. In *Kryptobaatar* the mediolateral diameter of the navicular is only slightly smaller than the maximal length of the astragalonavicular facet of the astragalus (1.5 mm and 1.8 mm, respectively); in *?Eucosmodon* the diameter of the navicular is much smaller than the length of the corresponding facet on the astragalus (3.8 mm and 5.5 mm, respectively). This demonstrates a wider range of the astragalonavicular movement in *?Eucosmodon* than in *Kryptobaatar*.

The dorsal surface of astragali is less undulate in Asian taxa than in *?Eucosmodon*. In relation to this the distal condyles on the tibia are more prominent in *?Eucosmodon* (Fig. 56E, F) than in *Kryptobaatar*, *Nemegtbaatar* (see 'Osteological descriptions') and an unidentified tibia from the Lance Formation (Fig. 55A).

It would be interesting to compare the lengths of metatarsals and digits in *Kryptobaatar* and *?Eucosmodon* and in extant small marsupial and eutherian mammals of different life-styles. In most scansorial mammals, e.g., *Glis glis*, *Tricho-*

surus vulpecula and *Phalanger maculatus* (Fig. 59), the fourth and fifth digits of the foot (but not the metatarsals) are the longest, which is an adaptation for holding branches. In *Didelphis marsupialis* elongation occurs but is less conspicuous. Similarly, in *Sciurus presicus*, which can move freely between the ground and trees (as, e.g., *Tupaia*; Jenkins 1974), the elongation of the phalanges of the fourth and fifth digits is modest and may be seen only in comparison with purely terrestrial forms belonging to the same family, such as, e.g., *Citellus xanthoprymnus* and *C. fulvus* (Table 5 and Fig. 59). In a terrestrial marsupial, *Antechinus stuarti*, the fourth and fifth digits are not elongated and are similar to those in *Citellus*, *Rattus* and other terrestrial rodents (Fig. 59).

Unfortunately, measurements of all the phalanges for all the digits cannot be taken for any of the three multituberculate genera (?*Eucosmodon*, *Ptilodus* and *Kryptobaatar*) in which incomplete pedes are known (Granger & Simpson 1929; Krause & Jenkins 1983; and this paper). In the case of ?*Eucosmodon* the association of individual phalanges with particular digits is not certain. In *Ptilodus* the digits II–V are completely preserved, and in *Kryptobaatar* digits IV and V are complete and in digits II and III the unguals have not been preserved. Because of this we compare below the lengths of the second to fifth digits without ungual phalanges of *Kryptobaatar*, with similarly incomplete digits of *Ptilodus* (rather than ?*Eucosmodon*) to make the comparisons more credible. In *Ptilodus* (after Krause & Jenkins 1983, Table 1) the combined length of Ph 1 and Ph 2 of D II is 12.1 mm, D III 11.9 mm, D IV 11.8 mm and D V 8.3 mm. In *Kryptobaatar* the combined lengths of Ph 1 and Ph 2 are: D II 8.3 mm, D III 7.9 mm, D IV 6.8 mm, D V 5.6. It follows that both taxa show similar pattern of relative lengths of foot digits (without ungual phalanges): DII > DIII > DIV > DV, the fifth being notably shorter than D II–D IV. This pattern differs from that characteristic of scansorial mammals (Fig. 59).

The lengths of the first ray cannot be compared, but it is important to note that while in ?*Eucosmodon* and in *Ptilodus kummae* Mt I is the shortest of all the metatarsals, in *Kryptobaatar* Mt V and Mt I are of the same length.

In Table 6 we give the lengths, widths and depths (the latter measured in dorsoplantar direction) of Mt I and the entocuneiform, as a percentage of the corresponding dimensions of MT III in multituberculates and small modern mammals of various life-styles. As may be seen in Table 6, Mt I in ?*Eucosmodon* is smaller in all dimensions than that in *Kryptobaatar*. Mt I in ?*Eucosmodon* is smaller in width and depth relative to Mt I in all modern mammals cited in Table 6, in particular in scansorial forms with an opposable hallux, as for example *Phalanger*, *Trichosurus*, *Marmosa* and *Didelphis*. The relative dimensions of Mt I in ?*Eucosmodon* are comparable (but a little smaller) to those of scansorial rodents, in which Mt I is not opposable, as for example *Glis* and *Sciurus*, and are similar to those of terrestrial rodents, such as *Rattus*, *Cricetulus* and *Citellus*, and to the terrestrial marsupial *Antechinus stuarti*.

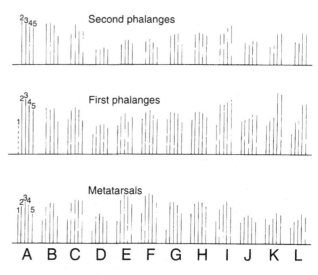

Fig. 59. Lengths of metatarsals of first and second phalanges in percent of total length of hind limb along third digit without ungual phalange in multituberculates and extant therian mammals. Phalanges enlarged ×2 in respect to the metatarsals. The metatarsals and phalanges are numbered. □A. *Kryptobaatar dashzevegi* (ZPAL MgM-I/41). □B. *Ptilodus kummae* (UA 9001, after Krause & Jenkins 1983, Table 1). □C. *Antechinus stuarti* (ZIN 13951). □D. *Mesocricetus raddei* (ZIN 660). □E. *Rattus norvegicus* (ZIN 83). □F. *Meriones tamariscinus* (ZIN 428). □G. *Citellus fulvus* (ZIN 175). □H. *Sciurus persicus* (ZIN 800). □I. *Glis glis* (ZIN 221). □J. *Didelphis marsupialis* (ZIN 66). □K. *Phalanger maculatus* (ZIN 10952). □L. *Trichosurus vulpecula* (ZIN 31713). C–G are terrestrial, H may move freely between the trees and ground, I–L are scansorial. Note elongation of phalanges of D IV and D V in scansorial forms. The respective lengths of the phalanges in multituberculates (A and B) are similar to those of terrestrial rather than scansorial taxa. (See also Table 5).

In three pedes of Mongolian multituberculates described here (right and left pedes of *Kryptobaatar* and left pes of *Chulsanbaatar*, Figs. 2, 6, 7, 25 and 54) the entocuneiform is elongated and protrudes strongly distally beyond the distal level of the mesocuneiform, ectocuneiform and the cuboid. The joint between the metatarsal I and entocuneiform is of a hinge type and permits movements of Mt I in a dorsoplantar direction; it seems, however, that a very small abduction was also possible. In mammals with an opposable hallux, this joint is saddle-shaped, and the entocuneiform does not protrude distally (or protrudes only slightly) beyond the level of the other cuneiforms. In these mammals the entocuneiform is wide relative to its length (Table 6) and the entocuneiform–metatarsal I joint is placed mediodistally, or medially (Lessertisseur & Saban 1967b; Starck 1979; Nowak & Paradiso 1983; Gebo *et al.* 1991; and many others), rather than distally as in Mongolian taxa. We conclude that the hallux was not divergent in *Kryptobaatar* and *Chulsanbaatar*. *Kryptobaatar* and *Chulsanbaatar* differ from ?*Eucosmodon* (Fig. 57) and *Ptilodus kummae* (Krause & Jenkins 1983, Fig. 27), where the entocuneiform protrudes less strongly distally beyond the level of the other cuneiforms. As far as the relative lengths of entocuneiforms are concerned, the entocunei-

Table 6. Length, width and depth of Mt I and entocuneiform in percent to the length, width and depth of Mt III.

Cat. No.	Species	MT I			Entocuneif.			H
		L	W	D	L	W	D	
UA 9001	*Ptilodus kummae*	58						?
ZPAL MgM -I/41	*Kryptobaatar dashzevegi*	73	71	112	33	168		t
AMNH 16325	*?Eucosmodon* sp.	63	70	58	29	116	108	?
ZIN 10952	*Phalanger maculata*	103	261	110	28	466	142	s
ZIN 31713	*Trichosurus vulpecula*	70	150	115	32	318	261	s
ZIN 1110	*Marmosa* sp.	86	180	76	12	316		s
ZIN 66	*Didelphis marsupialis*	72	131	110	14	258	194	s
ZIN 13951	*Antechinus stuarti*	57	86	100	26	114	100	t
ZIN 379	*Rattus norvegicus*	54	90	88	33	112	142	t
ZIN 221	*Glis glis*	55	92	69	39	110	187	s
ZIN 800	*Sciurus persicus*	77	83	107	20	100	214	s
ZIN 203	*Citellus xanthoprymnus*	62	71	100	30	130	137	t

Cat. No. = catalogue number; D = depth; H = habits; L = length; s = scansorial; t = terrestrial; W = width. The measurements of *Ptilodus kummae* are from Krause & Jenkins (1983, Table 1).

form in *?Eucosmodon* is, in fact, relatively longer than in *Kryptobaatar* (Table 6). As, however, the joint between the navicular and entocuneiform in *?Eucosmodon* is placed on the mediodistal side of the navicular, while in *Kryptobaatar* it is placed distally, the entocuneiform appears longer in *Kryptobaatar* than in *?Eucosmodon* (Figs. 54 and 57).

The distal facet on the entocuneiform in *?Eucosmodon* was referred to by Krause & Jenkins (1983) as saddle-shaped. These authors argued that the entocuneiform–Mt I joint permitted both flexion–extension and abduction–adduction of the hallux. The proximal part of the entocuneiform in *?Eucosmodon* is relatively narrower than in *Kryptobaatar* and possibly permitted a considerable degree of mobility (abduction) in the navicular–entocuneiform joint, apparently limited in *Kryptobaatar*.

The shape of the ungual phalanges of the *?Eucosmodon* foot speaks against its scansorial mode of life. In the scansorial mammals the ungual pedal phalanges are strongly compressed laterally and curved. Rose (1990), Van Valkenburgh (1987) and MacLeod & Rose (1993) studied the structure of ungual phalanges of the manus in Paleogene and some extant mammals. We were unable to make direct comparisons with their data, as the only multituberculate unguals available to us belong to the hind limb. The pedal unguals have been preserved in *?Eucosmodon* and *Kryptobaatar* (Fig. 2B), but the latter could not be prepared from the matrix. The two pedal ungual phalanges of *?Eucosmodon* (of undetermined digits) are wide and little curved and are reminiscent of those of modern terrestrial mammals such as *Citellus*, rather than those of scansorial taxa such as *Sciurus*, *Glis* and *Marmosa* (Fig. 60). As demonstrated by MacLeod & Rose (1993), the ungual phalanges of the manus differ in mammals of various life-styles: in scansorial forms they are deep and narrow, similar to the pedal ones of scansorial forms shown by us in Fig. 60C–E. The pedal ungual phalanx of *?Eucosmodon* sp. (Fig. 60A) is reminiscent of those of the manus of cursorial and some fossorial forms figured by MacLeod & Rose. However, the fossorial forms in question, e.g., *Cynomys ludovicianus* and *Marmota monax*, mainly use the incisors for digging rather than the phalanges (personal observations by PPG). In addition, MacLeod & Rose (1993) did not provide the transverse sections of the phalanges, and the comparisons of only the lateral and dorsal views may be misleading.

If follows from the foregoing comparisons that the pedes of *Kryptobaatar* and *Chulsanbaatar* differ in many characters from those of *?Eucosmodon* and *Ptilodus*. While arboreal adaptations of *?Eucosmodon* and *Ptilodus* might be possible, *Kryptobaatar* and *Chulsanbaatar* do not show such adaptations in their pedal structure. We conclude that *Kryptobaatar* and other Mongolian taxa studied by us were terrestrial runners.

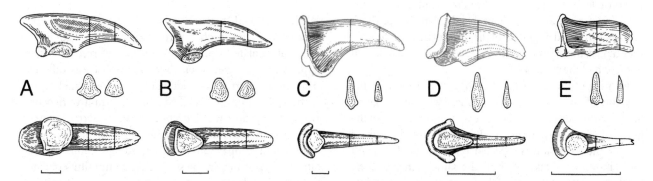

Fig. 60. Comparison of right ungual phalanges. Upper row – lateral view, middle – transverse sections along the lines shown in lateral and plantar views, bottom – plantar view. ☐A. *?Eucosmodon* sp. (AMNH 16325). ☐B. *Citellus xanthoprymnus* (ZIN 159). ☐C. *Sciurus presicus* (ZIN 800). ☐D. *Glis glis* (ZIN 221). ☐E. *Marmosa* sp. (ZIN 1110). Scale bars 2 mm.

Forelimb movements

In small mammals during fast asymmetrical gaits the forelimbs work mainly to absorb the shock of landing; the absorption increases with speed, steepness of the trajectory, and the animal's mass. We argued in the preceding sections that the multituberculate gait, when they were running fast, was asymmetrical and the trajectory of the jump was steeper than in extant therians of similar size and that because of this, they did not have the gathered phase of flight. If so, their forelimbs would act only for absorbing the shock of landing.

In frogs which usually also use asymmetrical locomotion, the coracosternal joint is retained. In frogs the short ribs do not join the sternum, and the muscles connecting the forelimb with the sternum and coracoid on one side and with the thoracic vertebrae on the other side act as shock absorbers (Gans 1961, 1974).

Eutherians, marsupials and, apparently independently, multituberculates solved the problem of absorption of the shock of landing by the reduction of the coracoid to a small process. The muscles of the shoulder girdle act as shock absorbers and move the thorax vertically between the shoulder girdle and forelimbs.

Cheng (1955) described migration of m. supracoracoideus onto the scapula and its division into m. supraspinatus and infraspinatus in the embryological development of *Didelphis*. In primitive tetrapods the main function of supracoracoideus is to keep the body from falling downwards between the limbs (Romer & Parsons 1986). The muscle fibers of supracoracoideus are arranged horizontally. In running lacertilians the forelimbs are only loaded slightly, and the horizontal arrangement of the muscle fibers is satisfactory to keep the body from sagging between the limbs.

At the absorption of the shock of landing during the asymmetrical gait, the main function of supracoracoideus increases, its fibers extend onto the scapula and are arranged vertically. This apparently was also the case in multituberculates which used asymmetrical gaits when running fast and in which supracoracoideus originated on the scapula. Romer & Parsons (1986, p. 296) stated that in mammals 'this major muscular migration [of supracoracoideus] is presumably responsible for the reduction of the coracoid region of the girdle'. We agree, and we believe that the demand for shock absorption caused migration of the fibers of this muscle onto the scapula, the disappearance of the coracosternal joint and reduction of the coracoid.

Twisting of the humerus (Fig. 46) is correlated with the position of the forelimbs. In animals with 'parasagittal' limbs, during the phase of flight, flexion and extension of the humeral and elbow joints occur; during the propulsive phase flexion and extension is repeated (Gambaryan 1974). All these movements take place in the same plane, and the parallel position of the axes of the humeral and elbow joints is most favorable. In mammals with abducted limbs (e.g., monotremes) the movements in the humeral joint occur in three planes arranged perpendicular to one another. The most important movements in this case are abduction, adduction and rotation, while flexion and extension are used to a smaller degree.

The origins of m. coracobrachialis and biceps brachii (see 'Myological reconstructions') allow speculations on the changes of these muscles at early stages of mammalian evolution. In cynodonts, as in modern reptiles, m. coracobrachialis longus, coracobrachialis brevis and biceps brachii were apparently present, all of them originating from the coracoid. During the reduction of the coracoid, all these origins came close to one another, producing an almost joint tendon (as was probably the case in multituberculates). In all reptiles and mammals m. biceps brachii inserts on the forearm, and the direction of its excursion does not interfere with the work of the shoulder joint. Musculus coracobrachialis longus inserts on the humeral entepicondyle and can work only if the line of its excursion does not interfere with the lesser tubercle, and coracobrachialis brevis inserts on the crest of the lesser tubercle. In multituberculates, because of the twisting of the humerus, the entepicondyle is situated more ventrally than in therians and the line of excursion of m. coracobrachialis longus and m. coracobrachialis brevis extended along the intertubercular groove (between greater and lesser tubercles). This groove in multituberculates is wider than in modern therians: in *Nemegtbaatar* ZPAL MgM-I/81 its width is 36% of the humeral width at the level of the tubercles, in *?Lambdopsalis* IVPP V8408 it is 36% and in IVPP V9051 37%. In rodents this groove occupies 17–23% of the humeral width. The width of the groove and the twisting of the humerus in multituberculates allowed the work of both coracobrachialis muscles and movements of the shoulder girdle.

Acquisition of the 'parasagittal' position of the forelimbs in therian mammals resulted in reduction of the humeral twisting. This eventually caused the medial shift of the coracoid process and a secondary division of the origin of m. coracobrachialis and m. biceps brachii.

Concluding remarks

We have argued that multituberculates had long spinous processes of the lumbar vertebrae and abducted limbs. Multituberculate locomotion is difficult to reconstruct, as apparently their gait when they moved fast was different from those occurring in modern mammals. Although asymmetrical, it cannot be classified either as primitive ricochet, ricochet, or gallop, which occur in small modern mammals (Gambaryan 1967, 1974). On the basis of the functional analysis and comparisons in the preceding sections, we tentatively suggest that in the studied Asian multituberculates the gait, when they were moving fast, was most similar to that of small extant mammals such as, e.g., *Meriones* (Gambaryan 1974; Gambaryan *et al.* 1978); the important difference was that the trajectory of the multituberculate jump was probably much steeper than in any modern small mammal. Abducted limbs and steep jumps limited multituberculate endurance for prolonged run.

Plesiomorphies and apomorphies of multituberculates

It is beyond the scope of this paper to provide a cladistic analysis of early mammals. However, the preceding comparisons of the skeleton of multituberculates with those of cynodonts and both non-therian and therian mammals, as well as the functional analysis, permit identification of plesiomorphic and apomorphic character states for multituberculates. This discussion also draws on previously published studies on the postcranial skeleton of multituberculates (Granger & Simpson 1929; Krause & Jenkins 1983; Szalay 1993; Sereno & McKenna 1990), as well as on studies on the postcranial skeleton of cynodonts and early mammals (Kühne 1956; Jenkins 1970b, 1971a; Jenkins & Parrington 1976; Sues 1985; Kemp 1980, 1982, 1983) and many others cited in the references.

Plesiomorphies

Vertebral column
- Lack of the transverse foramen in atlas.
- Cervical ribs.
- Lumbar vertebrae without anapophyses. Kühne (1986, Text-Fig. 45D) figured the apparent anapophyses on the lumbar vertebrae in *Oligokyphus*, but Sues (1985, and personal communication, March 1994) believes that they are lacking in tritylodontids. Jenkins (1970b, p. 235 wrote): 'Cynodont vertebrae typically bear anapophyses, short processes which represent the attachment of longissimus fascicles and which are common only in mammals'. Sues (1985) pointed out that the anapophyses are present in most, but not in all cynodonts.

Pectoral girdle
- Interclavicle and clavicle.

Forelimbs
- Abducted.

Humerus
- Teres tuberosity crescent-shaped.

- Radial and ulnar condyles (no trochlea).
- Entepicondylar foramen.
- Strong degree of twisting.

Pelvis
- Large iliosacral angle.

Hind limbs
- Abducted.

Tibia
- Mediolateral diameter greater than craniocaudal.

Synapomorphies

Vertebral column
- Axis with a well developed dens. Shared with tritylodontids and all mammals.

Ribcage
- Ossified sternebrae. Shared with tritylodontids and all mammals.

Shoulder girdle
- Scapulocoracoid with an incipient supraspinous fossa. Shared with *Cynognathus* (Gregory & Camp 1918) and tritylodontids (Kühne 1956, Fig. 52B), but lacking in Liassic triconodonts (Jenkins & Parrington 1976).
- Procoracoid lacking. Shared with therians. In tritylodontids and triconodonts the procoracoid is reduced in respect to cynodonts, but still present (Kühne 1956; Jenkins 1970b; Jenkins & Parrington 1976; Sues 1985).
- Prominent acromion. Shared with tritylodontids and all mammals, although in multituberculates and therians it is more prominent than in other mammals (Kühne 1956; Jenkins & Parrington 1976; Kemp 1983; and personal observations).

Humerus
- Lack of ectepicondylar foramen. Shared with tritylodontids and all mammals, but not with the monotremes. Jenkins (1970b, p. 238) stated that: 'There are basic similarities between a cynodont and monotreme humerus, for both retain large ect- and entepicondylar foramina'; however, we have not found the ectepicondylar foramen in *Ornithorhynchus*.

Ulna

- High, compressed olecranon process. Shared with tritylodontids and all mammals.

Pelvis

- Ilium rod-like, strongly elongated anteriorly, with a longitudinal ridge on the lateral side. Shared with tritylodontids and all mammals.
- Acetabulum open dorsocaudally. Shared with tritylodontids, triconodonts, some arboreal (gliding) marsupials and some swimming eutherians (*Desmana*).
- Lack of the posterior process of ilium. Shared with tritylodontids and all mammals.
- Epipubic bones. Shared with tritylodontids, monotremes, marsupials, possibly with earliest eutherians, unknown in triconodonts.

Femur

- Head spherical, on a constricted neck, set apart from the shaft. Shared with therians.
- Greater trochanter prominent, separated from the head by a deep incisure. Shared with therians.

Knee joint

- Patella. Shared with monotremes and therians, unknown in triconodonts, apparently absent in tritylodontids (Hans-Dieter Sues – personal communication, March 1994).
- Parafibula. Shared with most marsupials (Haines 1942).

Ankle joint

- Calcaneal tubercle high, laterally compressed. Shared with therians.

Autapomorphies:

Scapulocoracoid

- Coracoid developed as a small process on the ventral angle of the scapula and not bent medially as in therians.

Pelvis

- Very small ischial arc, placed high dorsally.
- Ischial tuber process-like.
- Postobturator foramen (or notch).
- Long ischiopubic symphysis, with ventral keel.
- Iliosacral contact dorsoventral.

Femur

- Lesser trochanter plate-like, convex lateroproximally, concave mediodistally and strongly protruding ventrally.
- Posttrochanteric fossa.
- Subtrochanteric tubercle.

Tibia

- Hook-like proximolateral process.

Fibula

- Hook-like proximolateral process.

Ankle joint

- Calcaneum with a deep peroneal groove.
- Peroneal tubercle clearly set off from the calcaneal body.
- Mediodistal calcaneocuboid facet.
- Calcaneo–Mt V contact.
- Astragalar canal (sulcus) surrounded by arched buttress (Szalay 1993).
- Astragalonavicular joint saddle-shaped.
- Abduction–adduction at the astragalonavicular joint, in a horizontal plane, around vertical axis.

The above synopsis, although it does not include the formal analysis of the distribution of character states, shows that there are ten postcranial skeleton synapomorphies shared by multituberculates and tritylodontids, the majority of which is also shared with all or some other mammals (see, however, Sues 1985 for criticism of tritylodontid–mammal relationships). Our synopsis does not support the conclusion of Rowe & Greenwald (1987) and Rowe (1988, 1993) on the close relationships of multituberculates and therians.

Rowe's definition of mammals has been criticized by Lucas (1990), Miao (1991) and Lucas & Luo (1993), while Wible (1991) and Miao (1993) challenged his cranial character analysis. To this criticism, with which we agree, we would like to add comments on the postcranial skeleton. Rowe (1988) recognized twenty postcranial-skeleton synapomorphies of the Theria and Multituberculata. We have found in the postcranial skeleton only four therian and multituberculate synapomorphies: procoracoid lacking; femoral head spherical, placed on a constricted neck, set apart from the shaft; greater trochanter prominent, separated from the head by a deep incisure, and high, laterally compressed, calcaneal tubercle. It seems that all of these characters developed independently in multituberculates and therians.

We give the list of eighteen autapomorphies that distinguish the multituberculate postcranial skeleton from those of all other mammals and cynodonts. It is, however, interesting that recent thorough analyses of the cranial structure of early mammals by Wible (1991), Wible & Hopson (1993),

Hopson & Rougier (1993), Rougier *et al.* (1993) and Lucas & Luo (1993) place the multituberculates closer to other mammals than the data from the postcranial skeleton do. One of the reasons may be that the braincase is more conservative than the postcranial skeleton and does not change easily with demands of the environment.

Another reason is that the characters cited by us as multituberculate autapomorphies were found in the Late Cretaceous and Early Tertiary multituberculates. As the postcranial skeletons of the Late Triassic, Jurassic and Early Cretaceous multituberculates are not known, it cannot be shown at which moment in multituberculate evolution these characters made their appearance. However, one can speculate that the multituberculate ankle joint could have developed from that of cynodonts (Jenkins 1971a), from which the Liassic triconodonts and therians can also be derived. These three mammal groups 'solved' the problems of functional anatomy each in its own way, and therefore it seems reasonable to presume that the functional adaptations unique to multituberculates were established at the beginning of their evolution. Our data agree with the conclusion of McKenna (1987) and Miao (1993) that multituberculates are a sister group to all the other mammals.

Habits and extinction

The habits of multituberculates have fascinated paleontologists since their discovery in the middle of the 19th century. Because of the superficial similarity of the lower jaw and the dentition pattern to rodents, they have been occasionally referred to as the rodents of the Mesozoic and generally regarded as the first mammals that occupied herbivorous niches. Krause (1982) reviewed the literature on multituberculate diets and on the basis of a detailed analysis of their dental function argued that they were omnivorous. As we have not investigated masticatory musculature in this paper (see Wall & Krause 1992 for biomechanical analysis of masticatory apparatus in *Ptilodus*), we shall not discuss the problem of multituberculate diet further.

As far as locomotion is concerned, Gidley (1909) suggested that Paleocene *Ptilodus* might have been saltatorial, Simpson (1926) suggested a possibility of semi-arboreal mode of life for multituberculates, and Simpson & Elftman (1928, p. 18) stated: 'The pelvic musculature, in agreement with all other known anatomical and environmental features, indicates an arboreal mode of life [of ?*Eucosmodon*]'. The question was further explored by Jenkins & Krause (1983) and Krause & Jenkins (1983), who argued that Paleocene *Ptilodus* and ?*Eucosmodon* were arboreal and concluded (p. 243) that 'some multituberculates, at least, were arboreal in habit'. Miao (1988) suggested that Paleocene–Eocene *Lambdopsalis* from China might have been a bur-

rower, a conclusion corroborated independently by Kielan-Jaworowska & Qi (1989).

The foregoing analysis of the postcranial anatomy of Late Cretaceous multituberculates from the Gobi Desert does not indicate arboreal habits for the studied forms. We believe that the habits of the multituberculates we studied were similar to those of modern murid rodents – jirds (known also as gerbils), in spite of notable, and sometimes even dramatic differences in anatomy of multituberculates and placentals in general. We have investigated, for comparative purposes, the musculature of two representatives of jirds *Meriones tamariscinus* and *Meriones blackleri*, both of which live in semi-desert regions of northern Caucasus and Armenia, respectively. *Meriones* species studied by us are approximately of *Nemegtbaatar* size. Although we have found that *Nemegtbaatar* and *Meriones* solved anatomical adaptations in a different way, their life-styles may have been comparable. Nowak & Paradiso (1983, p. 652) stated: 'Jirds inhabit clay and sandy deserts, bush country, and steppes, low plains, cultivated fields, grasslands and mountain valleys. They are terrestrial and construct burrows, where they spend much of their time.' *Meriones* species are nocturnal, which might have also been the case with studied multituberculates. We agree with Jerison (1973), Crompton *et al.* (1978) and many others that Mesozoic mammals were night dwellers (Fig. 61). Krause (1986) provided measurements of orbit size relative to skull size for two North American multituberculate species and argued that the multituberculates studied by him had eyes that were smaller than those of small-eyed, nocturnal living mammals. Krause (1986, p. 106) stated: 'Interestingly, while the eyes of *Ptilodus* are relatively small for living mammals, the olfactory bulbs are relatively the largest (Simpson, 1937; Jerison, 1973), suggesting that olfaction was indeed a dominant sense, more so than vision.' He concluded (contra Landry 1976) that multituberculates probably were nocturnal.

In all Late Cretaceous Asian multituberculates studied by us, the postorbital process is placed far posteriorly on the parietal (not on the frontal as in therian mammals), indicating the presence of big eyes. We argue elsewhere (Gambaryan & Kielan-Jaworowska, in preparation) that the recognition of small eyes by Krause (1986) in *Ptilodus* and *Ectypodus* was based on incorrect reconstructions of the position of the postorbital process (on the frontal) in these genera by Simpson (1937) and Sloan (1979). The presence of big eyes provides additional support to the idea of the nocturnal mode of life of multituberculates.

We speculate that, similarly to *Meriones*, Asian multituberculates inhabited semi-deserts. Our conclusions are based in part on the sedimentological evidence. Jerzykiewicz (1989) and Jerzykiewicz *et al.* (1994) argued that the dinosaur eggs, skeletons of dinosaurs, lizards, mammals and birds found in the Gobi Desert Djadokhta formation were buried in eolian sand. Kielan-Jaworowska (1977) claimed that there are no remnants of trees in the Djadokhta and Barun Goyot formations, although tree trunks are common in sandy dinosaur-bearing sediments of the younger Nemegt Formation, in which mammals have not been found. She concluded that Late Cretaceous mammals of the Djadokhta and Barun Goyot formations were living in steppe or semi-desert habitats. She also demonstrated that the two eutherian genera *Kennalestes* and *Asioryctes* found together with the multituberculates studied here had non-opposable pollex and hallux and certainly were not arboreal. The same holds for the somewhat larger eutherian genera *Zalambdalestes* and *Barunlestes* from the Djadokhta and Barun Goyot formations, respectively. These were compared in their habits to modern African Macroscelididae (Kielan-Jaworowska 1978), which also inhabit semi-desert regions.

Asian Late Cretaceous multituberculate genera do not show unequivocal fossorial adaptation, although a possibility of such adaptations has been suggested for an unidentified taeniolabidoid from the Djadokhta Formation (Kielan-Jaworowska 1989).

The structure of the pes of Asian multituberculates and ?*Eucosmodon* (and possibly also *Ptilodus*, which was not studied by us) indicates differences in life-styles. Some of the characters of the ?*Eucosmodon* pes, such as the high mobility of the tarsal joints, point to a scansorial mode of life, while others, such as relatively short fourth and fifth digits, gracile first metatarsal and wide, not laterally compressed ungual phalanges, speak against it. If ?*Eucosmodon* and other Paleocene multituberculates were indeed scansorial, they must have solved the adaptation to life in the trees somewhat differently than most modern mammals do.

We speculated under 'Functional anatomy' that the reduction of the coracoid took place in multituberculates as an adaptation for running with asymmetrical jumping. In the evolution of mammals the coracoid became reduced independently at least two (but possibly more) times: in multituberculates and in the common ancestor of marsupials and eutherians. We do not know whether it was reduced in the evolution of the triconodonts. In Liassic triconodonts, as demonstrated by Jenkins & Parrington (1976), the coracoid is relatively large, similar to that in cynodonts. The coracoid has not been preserved in the Early Cretaceous triconodont *Gobiconodon*, but the scapulae attributed to this taxon have a very large supraspinous fossa, similar to that in therians, which indicates a reduction of the coracoid (Jenkins & Schaff 1988, Fig. 13). Likewise, the coracoid has not been preserved in the only known postcranial skeleton of a docodont (Krusat 1991). The coracoid is reduced in an eupantotherian from the Late Jurassic of Portugal (Krebs 1991), and this may indicate that it was reduced in the common ancestor of marsupials and eutherians. We suggest that each time the coracoid became reduced, it was as an adaptation for asymmetrical jumping. If this is true, all of the mammals with reduced coracoids must have gone through a stage of the gait with asymmetrical jumping. Therefore we believe that the ancestors of mammals with a reduced coracoid did not pass

through an arboreal phase (contra Dollo 1899; Bensley 1901a, 1901b; Matthew 1904; Lewis 1964, 1989; Steiner 1965; Martin 1968; Rowe & Greenwald 1987; and many others; and in agreement with Gidley 1919; Haines 1958; Kielan-Jaworowska 1977; Szalay 1984; and (implicitly) with Napier 1961; Altner 1971; and Jenkins 1974).

The problem of multituberculate extinction has been thoroughly reviewed by Van Valen & Sloan (1966) who concluded that first condylarths, then primates, and finally rodents contributed to the gradual extinction of multituberculates. Hopson (1967) discussed the competitive inferiority of multituberculates to placental herbivores in the early Tertiary. Krause & Jenkins (1983, p. 244) stated: 'However attractive is Hopson's hypothesis, our review of postcranial skeleton of North American forms provides no evidence of features that might be considered significantly inferior to those of eutherians'. Krause (1986) reviewed the problem of competitive exclusion and taxonomic displacement in the Paleocene and Eocene mammal communities in North America. He demonstrated inverse correlations in generic diversity and relative abundance between multituberculates and rodents in the Paleocene and Eocene of the Western Interior of North America. He argued that the sudden appearance of paramyid rodents (which apparently originated in Asia) in earliest Clarforkian (latest Paleocene) deposits of North America (Gingerich & Rose 1977; Rose 1980, 1981) may have caused major restructuring of early Cenozoic mammal communities in North America, and that competitive exclusion may have played a role in the decline of multituberculates.

We agree with Hopson (1967) and Krause (1986). Although we are aware of the objection of Krause (1986) that competitive inferiority in terms of individual anatomical features cannot be demonstrated in the fossil record, we speculate that the structure of the pelvis and limb abduction may indicate competitive inferiority of multituberculates to placentals. As demonstrated by Kielan-Jaworowska (1979), the pelvis of the multituberculate *Kryptobaatar* shows a unique structure characterized by the complete fusion of the opposite pubes and ischia that form a ventral keel. We regard the ventral keel as a multituberculate autapomorphy. Kielan-Jaworowska (1979) argued that because of a very small ischial arc and a long immobile symphysis, the space available for the passage of an egg in *Kryptobaatar* would be less than 3.4 mm, which would be smaller than any known cleidoic egg (it would be still notably smaller in *Chulsanbaatar*). She concluded that multituberculates were born viviparous with an extremely small neonate.

Lillegraven (1975, 1979) argued that the appearance of the trophoblast and the ensuing prolongation of the gestation period had an enormous impact on the adaptative potentiality of eutherians. He stated (1979, p. 273): 'I believe that the extensive development of the eutherian trophoblastic function was unique and perhaps the single most important post-Jurassic event within the Mammalia. It allowed the eutheri-

Fig. 61. Reconstruction of *Nemegtbaatar*. Drawing by Bogdan Bocianowski.

ans to develop precocial young, enter new habitats, foster advanced societal behavior and ultimately make possible the evolution of culture and the historical record. Whole new realms of adaptation and behavior not possible within the marsupials became reality within the Eutheria because of this breakthrough in reproductive biology.' It is obvious that the extremely narrow pelvis and immobile symphysis of multituberculates precluded such evolutionary prospects.

Another character implying competitive inferiority of multituberculates to eutherians may be the abducted position of their limbs. The sprawling stance, as argued by Rewcastle (1981), is not adaptive for large and cursorial tetrapods: 'The shift to erect stance occurred in small tetrapods, initially as a cursorial adaptation, but also providing the possibility for dramatic size increase' (Rewcastle 1981, p. 262). Cursorial gait developed several times in tetrapod

evolution (in some dinosaurs, running birds, different hoofed mammals and others) and has always been associated with the upright limb posture. In multituberculates, because of the abducted limbs, the mechanics of jumps was very different from that in therian mammals. They apparently were able to run fast on short distances, using asymmetrical gaits with steep jumps, when passing open areas between the bushy parts of the semi-desert, but their endurance for running prolonged distances was limited. It is also possible that they moved relatively slowly, using symmetrical gaits, when seeking food in bushes around their burrows, as *Meriones* and many other small modern mammals also do (Gambaryan 1974).

We argued above that the keel in the multituberculate pelvis might have been developed as a response to the origin of femoral adductors ventral to the acetabulum; this in turn was related to the abducted position of the hind limbs. If this is true, one may speculate that the abducted position of multituberculate limbs would restrict their evolutionary possibilities also by limiting their reproductive strategy.

References

Altner, G. 1971: Histologische und vergleichend-anatomische Untersuchungen zur Ontogenie und Phylogenie des Handskeletts von *Tupaia glis* (Diard 1820) and *Micorcebus murinus* (J.F. Miller 1777). *Folia Primatologica, Supplement 14*, 1–106.

Archer, M., Flannery, T.F., Ritchie, A. & Molnar, R.E. 1985: First Mesozoic mammal from Australia – an early Cretaceous monotreme. *Nature 318*, 363–366.

Aristov, A.A., Krasts, I.V. & Gambaryan, P.P. 1980: Biomekhanika povorota bol'shego tushkanchika *Allactaga jaculus* Pall. [Biomechanics of the turning of a large gerboa *Allactaga jaculus* Pall.] *Trudy Zoologicheskogo Instituta Akademii Nauk SSSR 91*, 56–62.

Barnett, C.H. & Napier, J.R. 1953a: The rotary mobility of the fibula in eutherian mammals. *Journal of Anatomy, London 87*, 11–21.

Barnett, C.H. & Napier, J.R. 1953b: The form and mobility of the fibula in metatherian mammals. *Journal of Anatomy, London 67*, 207–213.

Bensley, B.A. 1901a: On the question of an arboreal ancestry of the Marsupialia and the interrelationships of the mammalian subclasses. *American Naturalist 35*, 117–138.

Bensley, B.A. 1901b: A theory on the origin and evolution of the Australian Marsupialia. *American Naturalist 35*, 245–269.

Benton, J.M. 1990: *Vertebrate Palaeontology: Biology and Evolution.* 377 pp. Unwin Hyman, London.

[Bleefeld, A.R. 1992: Functional Morphology in Paleontology: a Case Study of Therian Postcranial Elements from the Late Cretaceous Lance Formation, Wyoming. 222 pp. Unpublished Ph.D. thesis, University of Pennsylvania.]

Bonaparte, J.F. 1971: Los tetrapodos del sector superior de la formacion Los Colorados, La Rioja, Argentina (Triassico Superior). I Parte. *Opera Lilloana 22*, 1–183.

Broom, R. 1910: On *Tritylodon* and on the relationships of the Multituberculata. *Proceedings of the Zoological Society of London 50*, 760–768.

Broom, R. 1914: On the structure and affinities of the Multituberculata. *Bulletin of the American Museum of Natural History 33*, 115–134.

Brovar, V.Ya. 1935: Biomekhanika kholki (v svyazi s voprosom o roli ostistykh otrostkov u pozvonochnykh). [Biomechanics of the withers (in relation to the function of the spinous processes in vertebrates).] *Trudy Moskovskogo Zootekhnicheskogo Instituta 2*, 42–58.

Brovar, V.Ya. 1940: K analizu sootnoshenii mezhdu vesom golovy i dlinoi ostistykh otrostkov pozvonkov. [Analysis of the relationships between the weight of the head and the lengths of the spinous processes of the vertebrae]. *Archiv Anatomii, Gistologii i Embriologii 24*, 54–57.

Brown, J.C. & Yalden, D.W. 1973: The description of mammals – 2. Limbs and locomotion of terrestrial mammals. *Mammal review 3*, 107–134.

Bryant, H.N. & Seymour, K.L. 1990: Observations and comments on the reliability of muscle reconstruction in fossil vertebrates. *Journal of Morphology 206*, 109–117.

Carroll, R.L. 1976: *Noteosuchus* – the oldest known rhynchosaur. *Annals of the South African Museum 72*, 37–57.

Carroll, R.L. 1988: *Vertebrate Paleontology and Evolution.* 698 pp. Freeman, New York, N.Y.

Casamiquela, R.M. 1964: *Estudias Icnólogicos: Problemas y Métodos de la Icnologia con Aplicación al Estudio de Pisada Mesozoicas (Reptilia, Mammalia) de la Patagonia.* 229 pp. El Ministerio de Asuntos Sociales de la Provincia del Rio Negro. Buenos Aires.

Chatterjee, S. 1978: A primitive parasuchid (phytosaur) reptile from the Upper Triassic Maleri Formation of India. *Palaeontology 21*, 83–127.

Cheng, C.C. 1955: The development of the shoulder region in the opossum, *Didelphis virginiana*, with special reference to musculature. *Journal of Morphology 97*, 415–471.

Clemens, W.A., Jr. 1963: Fossil mammals of the type Lance Formation, Wyoming. Part I. Introduction and Multituberculata. *University of California Publications in Geological Sciences 48*, 1–105.

Clemens, W.A. & Kielan-Jaworowska, Z. 1979: Multituberculata. *In* Lillegraven, J.A., Kielan-Jaworowska, Z. & Clemens W.A. (eds.): *Mesozoic Mammals: the First Two-thirds of Mammalian History.* 99–149. University of California Press, Berkeley, Calif.

Crompton, A.W. & Jenkins, F.A. Jr. 1973: Mammals from reptiles: a review of mammalian origins. *Annual Review of Earth and Planetary Sciences 1*, 131–155.

Crompton, A.W., Taylor, C.R. & Jagger, J.A. 1978: Evolution of homeothermy in mammals. *Nature 272*, 333–336.

Cruickshank, A.R.I. 1972: The proterosuchian thecodonts. *In* Joysey, K.A. & Kemp, T.S. (eds.): *Studies in Mammalian Evolution.* 89–119. Oliver & Boyd, Edinburgh.

Davies, D.V. & Davies, F. 1962: *Gray's Anatomy, Descriptive and Applied.* 1632 pp. Longmans, Green & Co., London.

[Deischl, G.D. 1964: The Postcranial Anatomy of Cretaceous Multituberculate Mammals. 85 pp. Unpublished M.Sc. thesis, University of Minnesota, Minneapolis.]

Dobson, G.E. 1882–1990: *A Monograph of the Insectivora, Systematic and Anatomical. Parts I–III.* 172 pp. John Van Voorst, London.

Dollo, L. 1899: Les ancêtres des Marsupiaux étaient-ils arboricoles? *Traveaux de la Station Zoologique Wimereux 7*, 188–203.

Elftman, H.O. 1929: Functional adaptations of the pelvis in marsupials. *Bulletin of the American Museum of Natural History 58*, 189–232.

Evans, H.E. & Christensen, G.C. 1979: *Miller's Anatomy of the Dog.* 1181 pp. Saunders, Philadelphia, Pa.

Fokin, I.M. 1978: *Lokomotsiya i Morfologiya Organov Dvizheniya Tushkanchikov [Locomotion and Morphology of the Locomotory Organs in Jerboas].* 118 pp. Nauka, Moscow.

Fox, R.C. 1978: Upper Cretaceous terrestrial vertebrate stratigraphy of the Gobi Desert (Mongolian People's Republic and western North America). *In* Stelck, C.R. & Chatterton, D.E. (eds.): *Western and Arctic Canadian Biostratigraphy. Geological Association of Canada, Special Paper 18*, 577–594.

Gambarjan, I.S. 1990: Faktory evolyutsii lokomotornoy sistemy nizshikh tetrapod. [The evolutionary factors of the locomotor system in the lower Tetrapoda]. *Trudy Zoologicheskogo Instituta Akademii Nauk SSSR 215*, 9–37.

Gambaryan, P.P. 1960: *Prisposobitel'nye Osobennosti Organov Dvizheniya Royushchikh Mlekopitayushchikh. [The Adaptative Features of the Locomotory Organs in Fossorial Mammals].* 195 pp. Izdatel'stvo Akademii Nauk Armyanskoj SSR, Yerevan.

Gambaryan, P.P. 1967: Proiskhozhdenie mnogoobraziya allurov u mlekopitayushchikh. [The origin of the variety of gaits in mammals.] *Zhurnal Obshchej Biologii 28*, 289–305.

Gambaryan, P.P. 1974: *How Mammals Run.* 367 pp. Wiley, New York, N.Y. (Originally published in Russian in 1972).

Gambaryan, P.P., Pechenyuk, A.D. & Kartashova, T.M., Krasts, I.V., Zykova, L.J., Rukhjan, R.G. & Toporkova, T.M. 1978: Evolyutsiya asimmetrichnykh allurov u mlekopitayushchikh. [Evolution of the asymmetrical gaits in mammals]. *Trudy Zoologicheskogo Instituta Akademii Nauk SSSR 75*, 78–105.

Gans, C. 1961: A bullfrog and its prey. A look at the biomechanics of jumping. *Natural History 70*, February, 26–37.

Gans, C. 1974: *Biomechanics: An Approach to Vertebrate Biology.* 161 pp. The University of Michigan Press, Ann Arbor.

Gasc, J.P. 1967: Squelette hyobranchial. *In* Grassé, P.P. (ed.): *Traité de Zoologie 16:1*, 550–584. Masson et Cie, Paris.

Gebo, D.L., Dagosto, M. & Rose, K.D. 1991: Foot morphology and evolution in Early Eocene *Cantius. American Journal of Physical Anthropology 86*, 51–73.

Getty, R. 1975: *Sisson and Grossman's the Anatomy of the Domestic Animals.* 2 volumes. 2095 pp. Saunders, Philadelphia, Pa.

Gidley, J.W. 1909: Notes on the fossil mammalian genus *Ptilodus* with description of a new species. *Proceedings of the U.S. National Museum 36*, 611–626.

Gidley, J.W. 1919: Significance of divergence of the first digit in the primitive mammalian foot. *Journal of the Washington Academy of Sciences 9,* 273–280.

Gingerich, P.D. 1977: Patterns in evolution in the mammalian fossil record. In: Hallam, A. (ed.): *Patterns of Evolution,* pp. 469–500. Elsevier, Amsterdam.

Gingerich, P.D. 1984: Mammalian diversity and structure. *In* Broadhead, T.W. (ed.): *Mammals: Notes for a Short Course, University of Tennessee Studies in Geology 8,* 1–19.

Gingerich, P.D. & Rose, K.D. 1977: Preliminary report on the American Clark Fork mammal fauna, and its correlation with similar faunas in Europe and Asia. *Geobios, Mémoire Special 1,* 39–45.

Gradziński, R., Kielan-Jaworowska, Z. & Maryańska, T. 1977: Upper Cretaceous Djadokhta, Barun Goyot and Nemegt formations of Mongolia, including remarks on previous subdivisions. *Acta Geologica Polonica 27,* 281–318.

Granger, W. & Simpson, G.G. 1929: A revision of Tertiary Multituberculata. *Bulletin of the American Museum of Natural History 56,* 601–676.

Gray, J. 1968: *Animal Locomotion.* 479 pp. Weidenfeld & Nicolson, London.

Greene, E.G. 1935: Anatomy of the rat. *Transactions of the American Philosophical Society, N.S. 27,* 1–370.

Gregory, W.K. 1910: The orders of mammals. *Bulletin of the American Museum of Natural History 27,* 1–524.

Gregory, W.K. 1951: *Evolution Emerging,* Vol. 1, 736 pp., Vol. 2, 1013 pp. MacMillan, New York, N.Y.

Gregory, W.K. & Camp, C.L. 1918: Studies in comparative myology and osteology. III. *Bulletin of the American Museum of Natural History 38,* 447–563.

Hahn, G. 1973: Neue Zähne von Haramiyiden aus der Deutschen Ober-Trias und ihre Beziehungen zu den Multituberculaten. *Palaeontographica A 142,* 1–15.

Haines, R. W, 1942: The tetrapod knee joint. *Journal of Anatomy 76,* 270–301.

Haines, R.W. 1958: Arboreal or terrestrial ancestry of placental mammals. *Quarterly Review of Biology 33,* 1–23.

Henkel, S. & Krusat, G. 1980: Die Fossil-Lagerstätte in der Kohlengrube Guimarota (Portugal) und der erste Fund eines Docodontiden-Skelettes. *Berliner geowissenschaftliche Abhandlungen A 20,* 209–216.

Hildebrand, M. 1960: How mammals run. *Scientific American 202,* 148–157.

Hildebrand, M. 1988: *Analysis of Vertebrate Structure.* 701 pp. Wiley, New York, N.Y.

Hopson, J.A. 1967: Comments on the competitive inferiority of the multituberculates. *Systematic Zoology 16,* 352–355.

Hopson, J.A. & Barghusen, H.R. 1986: An analysis of therapsid relationships. *In* Hotton, N. III, MacLean, P.D., Roth, J.J. & Roth, E.C. (eds.): *The Ecology and Biology of Mammal-like Reptiles,* 83–106. Smithsonian Institution Press, Washington, D.C.

Hopson, J.A., Kielan-Jaworowska, Z. & Allin, E.F. 1989. The cryptic jugal of multituberculates. *Journal of Vertebrate Paleontology 9,* 201–209.

Hopson, J.A. & Rougier, G.W. 1993: Braincase structure in the oldest known skull of a therian mammal: implications for mammalian systematics and cranial evolution. *American Journal of Science 293A-A,* 268–299.

Howell, A.B. 1926: Anatomy of the wood rat. *Monographs of the American Society of Mammalogists 1,* 1–225.

Howell, A.B. 1933: The saltatorial rodent *Dipodomys*: the functional and comparative anatomy of its muscular and osseous systems. *Proceedings of the American Academy of Arts and Sciences 67,* 377–536.

Howell, A.B. 1944: *Speed in Animals. Their Specialization for Running and Leaping.* 270 pp. University of Chicago Press, Chicago, Ill.

Hurum, J.H. 1992. Earliest occurrence of sinus frontalis in Mammalia. Abstract. *20 Nordiska geologiska Vintermøtet, Reykjavik,* 77.

Hurum, J.H. 1994. The snout and orbit of Mongolian multituberculates studied by serial sections. *Acta Palaeontologica Polonica 39,* 181–221.

Huxley, T.H. 1879: On the characters of the pelvis in the Mammalia, and the conclusions respecting the origin of mammals which may be based on them. *Proceedings of the Royal Society of London 28,* 395–405.

International Commission on Veterinary Anatomical Nomenclature 1973: *Nomina Anatomica Veterinaria.* Adolf Holzhausen's Successors, Vienna. 218 pp.

Jenkins, F.A. Jr. 1970a: Limb movements in a monotreme (*Tachyglossus aculeatus*): a cineradiographic analysis. *Science 168,* 1473–1475.

Jenkins, F.A. Jr. 1970b: Cynodont postcranial anatomy and the 'prototherian' level of mammalian organization. *Evolution 24,* 230–252.

Jenkins, F.A. Jr. 1971a: The postcranial skeleton of African cynodonts. *Bulletin of the Peabody Museum of Natural History 36,* 1–216.

Jenkins, F.A. Jr. 1971b: Limb posture and locomotion in the Virginia opossum (*Didelphis marsupialis*) and in other non-cursorial mammals. *Journal of Zoology, London 165,* 303–315.

Jenkins, F.A. Jr. 1973: The functional anatomy and evolution of the mammalian humero–ulnar joint. *The American Journal of Anatomy 137,* 281–298.

Jenkins, F.A. Jr. 1974: Tree shrew locomotion and the origin of primate arborealism. *In* Jenkins, F.A. (ed.): *Primate Locomotion,* 85–116. Academic Press, New York, N.Y.

Jenkins, F.A. Jr. 1984: A survey of mammalian origins. *In* Broadhead, T.W. (ed.): *Mammals: Notes for a Short Course, University of Tennessee, Studies in Geology 8,* 32–47.

Jenkins, F.A., Jr. 1990: Monotremes and the biology of Mesozoic mammals. *Netherland Journal of Zoology 40,* 5–31.

Jenkins, F.A. Jr. & Camazine, S.C. 1977: Hip structure and locomotion in ambulatory and cursorial carnivores. *Journal of Zoology, London 181,* 351–370.

Jenkins, F.A., Jr. & Goslow, G.E. 1983: The functional anatomy of the shoulder of the savannah monitor lizard (*Varanus exanthematicus*). *Journal of Morphology 175,* 195–216.

Jenkins, F.A., Jr. & Krause, D.W. 1983: Adaptations for climbing in North American multituberculates (Mammalia). *Science 220,* 713–715.

Jenkins, F.A., Jr. & Parrington, F.R. 1976: Postcranial skeleton of the Triassic mammals *Eozostrodon, Megazostrodon,* and *Erythrotherium. Philosophical Transactions of the Royal Society of London B 273,* 387–431.

Jenkins, F.A. Jr. & Schaff, Ch.R. 1988: The Early Cretaceous mammal *Gobiconodon* (Mammalia, Triconodonta) from the Cloverly Formation in Montana. *Journal of Vertebrate Paleontology 6,* 1–24.

Jenkins, F.A. Jr. & Weijs, W.A. 1979: The functional anatomy of the shoulder in the Virginia opossum (*Didelphis virginiana*). *Journal of Zoology 188,* 379–410.

Jerison, H.J. 1973: *Evolution of the Brain and Intelligence,* 482 pp. Academic Press, New York, N.Y.

Jerzykiewicz, T. 1989: 1988 Sino–Canadian Dinosaur Project Expedition successful in Inner Mongolia. *Geos 18,* 1–6.

Jerzykiewicz, T., Koster, E.H. & Zheng, J.-J. 1993: Djadokhta Formation correlative strata in Chinese Inner Mongolia: an overview of the stratigraphy, sedimentary geology, and paleontology and comparisons with the type locality in the pre-Altai Gobi. *Canadian Journal of Earth Sciences 30,* 2180–2195.

Jouffroy, K.F. 1971: Musculature des membres. *In* Grassé, P. (ed.): *Traité de Zoologie,* Tome 16, Fascicule 2, *Mammifères, Musculature,* 1–476. Masson et Cie, Paris.

Jouffroy, F.K. & Lessertisseur, J. 1968: Musculature du tronc. *In* Grassé, P. (ed.): *Traité de Zoologie,* Tome 16, Fascicule 2, *Mammifères, Musculature,* 473–732. Masson et Cie, Paris.

Jouffroy, F.K. & Lessertisseur, J. 1971: Particularités musculaires des Monotrèmes. Musculature post-cranienne. *In* Grassé, P. (ed.): *Traité de Zoologie,* Tome 16, Fascicule 3, *Mammifères, Musculature,* 733–828. Masson et Cie, Paris.

Kemp, T.S. 1980: The primitive cynodont *Procynosuchus*: structure, function and evolution of the postcranial skeleton. *Philosophical Transactions of the Royal Society of London B 288,* 217–158.

Kemp, T.S. 1982: *Mammal-like Reptiles and the Origin of Mammals.* 363 pp. Academic Press, New York, N.Y.

Kemp, T.S. 1983: The relationships of mammals. *Zoological Journal of the Linnean Society 77,* 353–384.

Kielan-Jaworowska, Z. 1969: Discovery of a multituberculate marsupial bone. *Nature 222,* 1091–1092.

Kielan-Jaworowska, Z. 1970: New Upper Cretaceous multituberculate genera from Bayn Dzak, Gobi Desert. *Palaeontologia Polonica 21*, 35–49.

Kielan-Jaworowska, Z. 1971. Skull structure and affinities of the Multituberculata. *Palaeontologia Polonica, 25*, 5–41.

Kielan-Jaworowska, Z. 1974: Multituberculate succession in the Late Cretaceous of the Gobi Desert (Mongolia). *Palaeontologia Polonica 30*, 23–44.

Kielan-Jaworowska, Z. 1977: Evolution of the therian mammals in the Late Cretaceous of Asia. Part II. Postcranial skeleton in *Kennalestes* and *Asioryctes*. *Palaeontologia Polonica 37*, 65–83.

Kielan-Jaworowska, Z. 1978: Evolution of the therian mammals in the Late Cretaceous of Asia. Part III. Postcranial skeleton in Zalambdalestidae. *Palaeontologia Polonica 38*, 3–41.

Kielan-Jaworowska, Z. 1979: Pelvic structure and nature of reproduction in Multituberculata. *Nature 277*, 402–403.

Kielan-Jaworowska, Z. 1989: Postcranial skeleton of a Cretaceous multituberculate mammal. *Acta Palaeontologica Polonica 34*, 75–85.

Kielan-Jaworowska, Z. 1992: Interrelationships of Mesozoic mammals. *Historical Biology 6*, 185–202.

Kielan-Jaworowska, Z. 1994: A new generic name for the multituberculate mammal 'Djadochtatherium' catopsaloides. *Acta Palaeontologica Polonica 39*, 134–136.

Kielan-Jaworowska, Z., Crompton, A.W. & Jenkins, F.A. Jr. 1987: The origin of egg-laying mammals. *Nature 326*, 871–873.

Kielan-Jaworowska, Z. & Dashzeveg, D. 1978: New Late Cretaceous mammal locality in Mongolia and a description of a new multituberculate. *Acta Palaeontologica Polonica 23*, 115–130.

Kielan-Jaworowska, Z. & Nessov, L.A. 1992: Multituberculate mammals from the Cretaceous of Uzbekistan. *Acta Palaeontologica Polonica 37*, 1–17.

Kielan-Jaworowska, Z., Presley, R. & Poplin, C. 1986: The cranial vascular system in taeniolabidoid multituberculate mammals. *Philosophical Transactions of the Royal Society of London B 313*, 525–602.

Kielan-Jaworowska, Z. & Qi T. 1990: Fossorial adaptations of a taeniolabidoid multituberculate mammal from the Eocene of China. *Vertebrata PalAsiatica 28*, 81–94.

Kielan-Jaworowska, Z. & Sloan, R.E. 1979: *Catopsalis* (Multituberculata) from Asia and North America and the problem of taeniolabidoid dispersal in the Late Cretaceous. *Acta Palaeontologica Polonica 24*, 187–197.

Krause, D.W. 1982: Jaw movement, dental function, and diet in the Paleocene multituberculate *Ptilodus*. *Paleobiology 8*, 265–281.

Krause, D.W. 1986: Competitive exclusion and taxonomic displacement in the fossil record: the case of rodents and multituberculates in North America. *In* Flanagan, K.M. & Lillegraven, J.A. (eds.): *Vertebrates, Phylogeny and Philosophy. Contributions to Geology, University of Wyoming, Special Paper 3*, 95–117.

Krause, D.W. & Baird, D. 1979: Late Cretaceous mammals east of the North American Western Interior Seaway. *Journal of Paleontology 53*, 562–565.

Krause, D.W. & Jenkins, F.A. Jr. 1983: The postcranial skeleton of North American multituberculates. *Bulletin of the Museum of Comparative Zoology 150*, 199–246.

Krebs, B. 1991: Das Skelett von *Henkelotherium guimarotae* gen et. sp. nov. (Eupantotheria, Mammalia) aus dem Oberen Jura von Portugal. *Berliner Geowissenschaftliche Abhandlungen A, 133*, 1–121.

Krishtalka, L., Emry, R.J., Storer, J.E. & Sutton, J.F. 1982: Oligocene multituberculates (Mammalia: Allotheria): youngest known record. *Journal of Paleontology 56*, 791–794.

Krusat, G. 1991: Functional morphology of *Haldanodon expectatus* (Mammalia, Docodonta) from the Upper Jurassic of Portugal. *In* Kielan-Jaworowska, Z., Heintz, N. & Nakrem, H.A. (eds.): *Fifth Symposium on Mesozoic Terrestrial Ecosystems and Biota. Extended Abstracts. Contributions from the Paleontological Museum, University of Oslo 364*, 37–38.

Kühne, W.B. 1956: *The Liassic Therapsid Oligokyphus.* 149 pp. British Museum (Natural History), London.

Kummer, B. 1959a: *Bauprinzipien des Säugerskeletts.* 235 pp. Georg Thieme, Stuttgart.

Kummer, B. 1959b: Biomechanik des Säugetierskeletts 6 (2). *In* Helmcke, J.-G., Lengerken, H. von & Starck, D. (eds.): *Handbuch der Zoologie, 8 Band, 24 Lieferung*, 1–80. Walter de Gruyter, Berlin.

[Kuznetsov, A.N. 1983: K Izucheniyu Lokomotsii i Lokomotornogo Apparata Seroj Krysy (*Rattus norvegicus*). [Study of Locomotion and Locomotory Apparatus of the Rat (*Rattus norvegicus*)]. 53 pp. Unpublished M. Sc. thesis, Gosudarstvennyi Universityet im. M.V. Lomonosova, Moscow.]

Kuznetsov, A.N. 1985: Sravnitel'no-funktsionalnyj analiz perednikh i zadnikh konechnostej mlekopitayushchikh. [Comparative-functional analysis of the fore- and hindlimbs in mammals]. *Zoologicheskii Zhurnal 64*, 1862–1867.

Landry, S.O. 1976: Disappearance of multituberculates. *Systematic Zoology 16*, 172–173.

Le Damany, P. 1906: L'angle sacropelvien. *Journal de l'anatomie et de la physiologie normales et pathologiques de l'homme et des animaux 42*, 153–192.

Leonardi, G., 1987: *Glossary and Manual of Tetrapod Footprint Palaeoichnology.* 75 pp. Ministerio das Minas e Energia, Brasília.

Lessertisseur, J. 1967: L'angle iliosacré des reptiles aux mammifères, son interprétation, son intéret paléontologique. *Colloques Internationaux du Centre National de la Recherche Scientifique 218*, 475–481.

Lessertisseur, J. & Saban, R. 1967a: Squelette axial. *In* Grassé, P.P. (ed.): *Traité de Zoologie*, Tome 16, Fascicule 1, *Mammifères, Téguments et Squelette*, 585–708. Masson et Cie, Paris.

Lessertisseur, J. & Saban, R. 1967b: Squelette appendiculaire. *In* Grassé, P.P. (ed.): *Traité de Zoologie*, Tome 16, Fascicule 1, *Mammifères, Téguments et Squelette*, 709–1078. Masson et Cie, Paris.

Lewis, O.J. 1963: The monotreme cruro–pedal flexor musculature. *Journal of Anatomy, London 97*, 55–63.

Lewis, O.J. 1964: The evolution of the long flexor muscles of the leg and foot. *International Review of General and Experimental Zoology 1*, 165–185.

Lewis, O.J. 1989. *Functional Morphology of the Evolving Hand and Foot.* 359 pp. Clarendon, Oxford.

Lillegraven, J.A. 1975: Biological considerations of the marsupial–placental dichotomy. *Evolution 29*, 707–722.

Lillegraven, J.A. 1979: Reproduction in Mesozoic mammals. *In* Lillegraven, J.A., Kielan-Jaworowska, Z. & Clemens, W.A. (eds.): *Mesozoic Mammals: The First Two-thirds of Mammalian History*, 259–276. University of California Press, Berkeley, Calif.

Lillegraven, J.A., Kielan-Jaworowska, Z. & Clemens, W.A. 1979: *Mesozoic Mammals: The First Two-thirds of Mammalian History.* 311 pp. University of California Press, Berkeley, Calif.

Lillegraven, J.A. & McKenna, M.C. 1986: Fossil mammals from the 'Mesaverde' Formation (Late Cretaceous, Judithian) of the Bighorn and Wind River Basins, Wyoming, with definitions of Late Cretaceous North American land-mammal 'Ages'. *American Museum Novitates 2840*, 1–68.

Lucas, S.G. 1990. The extinction criterion and the definition of the Class Mammalia. *Journal of Vertebrate Paleontology 8, Supplement to No. 3, Abstracts*, 33A.

Lucas, S.G. & Luo, Z. 1993. *Adelobasileus* from the Upper Triassic of West Texas: the oldest mammal. *Journal of Vertebrate Paleontology 13*, 309–334.

MacLeod, N. & Rose, K.D. 1993: Inferring locomotor behavior in Paleogene mammals via eigenshape analysis. *American Journal of Science 293, A*, 300–355.

McKenna, M.C. 1961: On the shoulder girdle of the mammalian subclass Allotheria. *American Museum Novitates 2066*, 1–27.

McKenna, M.C. 1975: Toward a phylogenetic classification of the Mammalia. *In* W.P. Luckett & F.S. Szalay (eds.): *Phylogeny of the Primates*, 21–46. Plenum, New York, N.Y.

McKenna, M.C. 1987: Molecular and morphological analysis of high-level mammalian interrelationships. *In* Patterson, C. (ed.): *Molecules and Morphology in Evolution: Conflict or Compromise?*, 55–93. Cambridge University Press, Cambridge.

Martin, R.D. 1968: Towards a new definition of primates. *Man 3*, 377–401.

Matthew, W.D. 1904: The arboreal ancestry of the Mammalia. *American Naturalist 38*, 811–818.

Miao, D. 1988: Skull morphology of *Lambdopsalis bulla* (Mammalia, Multituberculata) and its implications to mammalian evolution. *Contributions to Geology, University of Wyoming, Special Paper 4*, 1–104.

Miao, D. 1991: On the origins of mammals. *In* Schultz, H.P. & Trueb, L. (eds.): *Origins of the Higher Groups of Tetrapods: Controversy and Consensus*, 579–597. Cornell University Press, Ithaca, N.Y.

Miao, D. 1993: Cranial morphology and multituberculate relationships. *In* Szalay, F.S., Novacek, M.J. & McKenna, M.C. (eds.): *Mammal Phylogeny: Mesozoic Differentiation, Multituberculates, Monotremes, Early Therians, and Marsupials*, 63–74. Springer, New York, N.Y.

Muybridge, E. 1957: *Animals in Motion* (edited by L.S. Brown). 74 pp. Dover, New York, N.Y.

Napier, J.R. 1961: Prehensibility and opposability in the hands of primates. *Symposium of the Zoological Society 5*, 115–132.

Novacek, M.J. 1992: Mammalian phylogeny: shaking the tree. *Nature 356*, 121–125.

Nowak, R.M. & Paradiso, J.L. 1983: *Walker's Mammals of the World*, Vols. I and II. 1362 pp. The Johns Hopkins University Press, Baltimore, Md.

Parrington, F.R. 1961: The evolution of mammalian femur. *Proceedings of the Zoological Society of London 137*, 285–298.

Pridmore, P.A. 1985: Terrestrial locomotion in monotremes (Mammalia, Monotremata). *Journal of Zoology, London 205*, 53–73.

Rewcastle, S.C. 1981: Stance and gait in tetrapods: an evolutionary scenario. *In* Day, M.H. (ed.): *Vertebrate Locomotion*, 239–268. The Zoological Society of London & Academic Press, London.

Rinker, G.C. 1954: The comparative myology of the mammalian genera *Sigmodon, Oryzomys, Neotoma* and *Paromyscus* (Cirtecinae), with remarks on intergeneric relationships. *Miscellaneous Publications of the Museum of Zoology, University of Michigan 83*, 1–124.

Romer, A.S. 1956: *Osteology of the Reptiles*. 772 pp. The University of Chicago Press, Chicago, Ill.

Romer, A.S. 1966: *Vertebrate Paleontology*. 468 pp. The University of Chicago Press, Chicago, Ill.

Romer, A.S. & Parsons, T.S. 1986: *The Vertebrate Body*. 679 pp. Sounders College Publishing, Philadelphia, Pa.

Rose, K.D. 1980: Clarkforkian Land-Mammal Age: revised definition, zonation, and tentative intercontinental correlations. *Science 208*, 744–746.

Rose, D.K. 1981: The Clarkforkian Land-Mammal Age and mammalian faunal composition across the Paleocene–Eocene boundary. *University of Michigan Papers on Paleontology 26*, 1–196.

Rose, K.D. 1990: Postcranial skeletal remains and adaptations in early Eocene mammals from the Willwood Formation, Bighorn Basin, Wyoming. *Geological Society of America, Special Paper 243*, 107–133.

Rougier, G.W., Wible, J.R. & Hopson, J.A. 1992: Reconstruction of the cranial vessels in the Early Cretaceous mammal *Vincelestes neuquenianus*: implications for the evolution of mammalian cranial system. *Journal of Vertebrate Paleontology 12*, 188–216.

Rowe, T. 1988: Definition, diagnosis, and origin of Mammalia. *Journal of Vertebrate Paleontology 8*, 241–264.

Rowe, T. 1993: Phylogenetic systematics and the early history of mammals. *In* Szalay, F.S., Novacek, M.J. & McKenna, M.C. (eds.): *Mammal Phylogeny: Mesozoic Differentiation, Multituberculates, Monotremes, Early Therians, and Marsupials*, 129–145. Springer, New York, N.Y.

Rowe, T. & Greenwald, N.S. 1987: The phylogenetic position and origin of Multituberculata. *Journal of Vertebrate Paleontology 7, Supplement to No. 3, Abstracts*, 24A–25A.

Sahni, A. 1972: The vertebrate fauna of the Judith River Formation, Montana. *Bulletin of the American Museum, Natural History 147*, 323–412.

Schaeffer, B. 1941a: The pes of *Bauria cynops* Broom. *American Museum Novitates 1103*, 1–7.

Schaeffer, B. 1941b: The morphological and functional evolution of the tarsus in amphibians and reptiles. *Bulletin of the American Museum of Natural History 78*, 395–472.

Schaller, O. (ed.) 1992: *Illustrated Veterinary Anatomical Nomenclature*. 614 pp. Ferdinand Enke, Stuttgart.

Sereno, P.C. & McKenna, M.C. 1990: The multituberculate clavicle and interclavicle, and the early evolution of the mammalian pectoral girdle. *Journal of Vertebrate Paleontology 10, Supplement to No. 3, Abstracts*, 42A.

Sigogneau-Russell, D. 1989: Haramiyidae (Mammalia, Allotheria) en provenance du Trias supérieur de Lorraine (France). *Palaeontographica A 206*, 137–198.

Simpson, G.G. 1926: Mesozoic Mammalia. IV. The multituberculates as living mammals. *American Journal of Science 2*, 228–250.

Simpson, G.G. 1928a: Further notes on Mongolian Cretaceous mammals. *American Museum Novitates 329*, 1–14.

Simpson, G.G. 1928b: *A Catalogue of the Mesozoic Mammalia in the Geological Department of the British Museum*. 215 pp. British Museum (Natural History), London.

Simpson, G.G. 1933. The 'plagiaulacoid' type of mammalian dentition. *Journal of Mammalogy 14*, 97–107.

Simpson, G.G. 1937: Skull structure of the Multituberculata. *Bulletin of the American Museum of Natural History 73*, 727–763.

Simpson, G.G. & Elftman, H.O. 1928: Hind limb musculature and habits of a Paleocene multituberculate. *American Museum Novitates 333*, 1–19.

Slijper, E.J. 1946: Comparative biologic–anatomical investigations on the vertebral column and spinal musculature of mammals. *Verhandlingen der Koninklijke Nederlandsche Akademie van Wetenschappen, Afd. Natuurkunde, (Tweede Sectie) 42:5*, 1–128.

Sloan, R.E. 1979. Multituberculata. *In* Fairbridge, R.W. & Jablonski, D. (eds.): *The Encyclopaedia of Paleontology*, 492–498. Gowden, Hutchison & Ross, Stroudsburg.

Sloan, R.E. & Van Valen, L. 1965: Cretaceous mammals from Montana. *Science 148*, 220–227.

Sokolov, I.I., Sokolov, A.S. & Klebanova E.A. 1974: Morfologicheskie osobennosti organov dvizheniya nekotorykh kun'ikh (Mustellidae) v svyazi s obrazom zhizni. [Morphological characters of the locomotory organs in some Mustellidae, in relation to their habits]. *Trudy Zoologicheskogo Instituta Akademii Nauk SSSR 54*, 4–98.

Starck, D. 1979: *Vergleichende Anatomie der Wirbeltiere auf evolutionsbiologischer Grundlage*. Bd. 2. 776 pp. Springer, Berlin.

Steiner, H. 1965: Die vergleichend-anatomische und ökologische Bedeutung der rudimentären Anlage eines selbständigen fünften Carpale bei *Tupaia*. Betrachtungen zum Homologie-Problem. *Israel Journal of Zoology 14*, 221–233.

Sues, H.-D. 1985: The relationships of the Tritylodontidae (Synapsida). *Zoological Journal of the Linnean Society 85*, 205–217.

Sukhanov, V.B. 1974: *Symmetrical Locomotion of Terrestrial Vertebrates and Some Features of Movement of Lower Tetrapods*. 274 pp. Amerind Publishing Co., New Delhi, Bombay, Calcutta, New York. (Originally published in Russian in 1968).

Swisher, C.C. III & Prothero, D.R. 1990: Single-Crystal $^{40}Ar/^{39}Ar$ dating of the Eocene–Oligocene transition in North America. *Science 249*, 760–762.

Szalay, F.S. 1977: Phylogenetic relationships and a classification of the Eutherian Mammalia. *In* Hecht, M.K., Goody, P.C. & Hecht, B.M. (eds.): *Major Patterns of Vertebrate Evolution*, 315–474. *Nato Advanced Study Institute, Ser. A:14*, Plenum, New York, N.Y.

Szalay, F.S. 1984: Arboreality: is it homologous in metatherian and eutherian mammals? *In* Hecht, M.K., Wallace, B. & Prance G.T. (eds.): *Eolutionary Biology 18*, 215–258. Plenum, New York, N.Y.

Szalay, F.S. 1990: Evolution of the tarsal complex in Mesozoic mammals. *Journal of Vertebrate Paleontology 10, Supplement 3*, p. 45A.

Szalay, F.S. 1993: Pedal evolution of mammals in the Mesozoic: tests for taxic relationships. *In* Szalay, F.S., Novacek, M.J. & McKenna, M.C. (eds.): *Mammal Phylogeny: Mesozoic Differentiation, Multituberculates, Monotremes, Early Therians, and Marsupials*, 108–128. Springer, New York, N.Y.

Szalay, F.S. & Decker, R.L. 1974: Origins, evolution and function of tarsus in the Late Cretaceous Eutheria and Paleocene Primates. *In* Jenkins, F.A. (ed.): *Primate Locomotion*, 233–259. Academic Press, New York, N.Y.

Turnbull, W.D. 1970: Mammalian masticatory apparatus. *Fieldiana: Geology 18*, 1–356.

Van Valen, L. & Sloan, R.E. 1966: The extinction of multituberculates. *Systematic Zoology 15*, 261–278.

Van Valkenburgh, B. 1987: Skeletal indicators of locomotor behavior in living and extinct carnivores. *Journal of Vertebrate Paleontology 7*, 162–182.

Verheyen, W.N. 1961: Recherches anatomiques sur *Micropotamogale ruwenzorii. Bulletin de la Societé Royal de Zoologie, Anvers 21*, 1–27, *22*, 1–28.

Vialleton, L. 1924. *Membres et Ceintures des Vertébrés Tétrapodes.* 710 pp. Librairie Octave Doin, Paris.

Wall, C.E. & Krause, W.D. 1992: A biomechanical analysis of the mastica-tory apparatus of *Ptilodus* (Multituberculata). *Journal of Vertebrate Paleontology 12*, 172–187.

Wible, J.R. 1991: Origin of Mammalia: the craniodental evidence reexam-ined. *Journal of Vertebrate Paleontology 11*, 1–28.

Wible, J.R. & Hopson, J. 1993: Basicranial evidence for early mammal phylogeny. *In* Szalay, F.S., Novacek, M.J. & McKenna, M.C. (eds.): *Mammal Phylogeny: Mesozoic Differentiation, Multituberculates, Mono-tremes, Early Therians, and Marsupials,* 45–62. Springer, New York, N.Y.

Zug, G.R. 1972: Anuran locomotion: structure and function. I. Preliminary observations on relation between jumping and osteometrics of appen-dicular and postaxial akeleton. *Copeia 1972*, 613–624.